Designing
Map Interfaces

Designing **Map Interfaces**

Patterns for Building Effective Map Apps

Michael Gaigg

Esri Press
REDLANDS | CALIFORNIA

Esri Press, 380 New York Street, Redlands, California 92373-8100

Printed in the United States of America.

ISBN: 9781589487253
Library of Congress Control Number: 2022946794

For purchasing and distribution options (both domestic and international), please visit esripress.esri.com.

To my wife, Jessica:

You are my sunshine! Your love inspires me to become a better person every day. And yes, you have better taste than I do (most of the time).

To my children, Sebastian, Nikolas, and Benjamin:

I can't wait to see all the wonderful and creative things you will accomplish in your lives.

To my mum and in loving memory of my dad:

Thank you for everything you've done for me.

Contents

Preface

The purpose of the book *Designing Map Interfaces: Patterns for Building Effective Map Apps* is to explain principles and best practices for planning and building map apps. Whether you are configuring an out-of-the-box solution, building an app using one of the readily available app builder tools, or working on a custom app project, this book will guide you toward developing more useful and usable apps. Our goal is to provide you with the know-how that will turn your next project into a success story.

What are UI patterns?

The user interface (UI) is the visual part of an app that determines how a user interacts with it and how information is displayed on the screen. To understand, plan, and build a useful and usable UI, you need a language that describes best practices. The main building blocks of this language are UI patterns, or *patterns* for short. They describe solutions that were distilled from recurring problems observed over the years. For convenience and clarity, each pattern described in the book follows the same structure:

- Name of the pattern
- Picture of the pattern showing an archetypal example
- What problem the pattern aims to solve
- Why the pattern is important
- When you should use this pattern
- How the pattern should be implemented
- Example of an implementation in the real world

Throughout the book, patterns are discussed in a general and abstract way because there is never a one-size-fits-all solution to a problem. On the contrary, a pattern is meant to assist and encourage you to build on, to improve, to use as a starting point, and to iterate until it meets your specific needs and conditions. Sometimes it may feel tempting to add this one little feature to an app, but only the knowledge of the correct usage and understanding of its implications will help you make an educated decision as to what, why, when, and how to solve your problem.

How to use this book

The book is grouped into seven chapters. Each chapter has an introduction that covers the concepts followed by a list of related patterns. The first and last chapters talk about general concepts and pitfalls, whereas the chapters in between cover more in-depth topics, as follows:

- **Chapter 1** introduces you to the design process, explains the importance of defining the user experience (UX) strategy, wireframes, and mock-ups. It describes design principles and the benefits of using a design system.
- **Chapter 2** helps you identify your target persona and their preferred app type. This knowledge will directly influence your choice of layout.
- **Chapter 3** discusses navigation and information retrieval when interacting with the map.
- **Chapter 4** explains solution strategies for dealing with complex data.
- **Chapter 5** describes the challenges and opportunities when designing for mobile devices and outlines common mobile workflows.
- **Chapter 6** goes into the details of single-purpose apps and their specific characteristics.
- **Chapter 7** lists common mistakes observed in map apps and how to avoid them or solve the problem.

The collection of patterns is a catalog that doesn't need to be read in linear fashion. You may find yourself skipping patterns and jumping right to what is important to you. Patterns also vary in size; some are unique and "atomic," whereas others are composed of multiple patterns. As a result, patterns described earlier in the book may refer to patterns in later chapters and vice versa. You are encouraged to find your own path for studying the content. In the book, patterns are marked in **green**.

Terminology

Throughout the book, some terms are used broadly to simplify the reading experience. Following is a list of common terms:

- **App** is a website, mobile responsive website, or native mobile app.
- **User** is a person using and interacting with an app.
- **Map** is an interactive map.
- **Extent** is the limit of the geographic area shown in the map.
- **Scale** is the ratio of a distance on the map to the corresponding distance on the ground. In an app, scale refers to the zoom level. For instance, a map at small scale is "zoomed out," whereas a map at large scale is "zoomed in."
- **Interface** references the UI and describes the space in which a user will interact with the computer, usually through a set of UI elements such as buttons, links, or a map.
- **Component** is the equivalent of a UI element and can be a simple element such as a button or a composite component such as a card that includes other simple components.
- **Click** is used interchangeably for a mouse click or a tap on a component through finger gestures.
- **Builder** is a person designing, developing, or configuring an app interface, including designers, developers, and configuration specialists.

List of patterns

The following map-related UI patterns are available. The patterns deemed most relevant to app developers are included in the book, but because of limited space, the remaining patterns are referenced for further reading using external links. The patterns are listed in alphabetical order here with the main chapter cited for each, although the patterns may be discussed elsewhere throughout the book.

- **2D** (external)
- **3D** (3)
- **Attribute filter** (4)
- **Augmented reality** (5)
- **Basemap** (external)
- **Basemap gallery** (external)
- **Basemap toggle** (external)
- **Blue dot** (5)

- Browse geographies (external)
- Buffer (external)
- Callout (external)
- Chart marker (4)
- Choropleth (external)
- Cluster marker (4)
- Data dimming (external)
- Data extract (external)
- Data–ink ratio (7)
- Desert fog (7)
- Embedded map (2)
- Empty state (1)
- Extent-driven content (6)
- Feature selection (4)
- Floor selector (5)
- Flow map (external)
- Focal point (1)
- Focus on the subject (2)
- Full map (2)
- Heat map (external)
- Home button (external)
- Info panel (3)
- Info pop-up (3)
- Kiosk map (external)
- Kitchen sink (7)
- Landing page (1)
- Layer list (external)
- Legend (external)
- List and details (6)
- Locate me (5)
- Location finder (1)
- Location list (6)
- MapTip (3)
- Marker (3)
- Meet in the middle (external)
- Nearby (external)
- No map (2)
- Offline maps (5)
- Onboarding (external)
- Overview map (external)
- Partial map (2)
- Placemarks (external)
- Progress indicator (external)
- Reference map (2)
- Rich marker (3)
- Route directions (external)
- Route positioning (external)
- Scroll to pan (6)
- Search (1)
- Search and edit (6)
- Search this area (5)
- Share live location (external)

- Side by side (external)
- Small multiples (external)
- Spark maps (external)
- Spatial filter (4)
- Store locator (6)
- Swipe (external)
- Task oriented (1)
- Theme toggle (external)
- Timeline slider (4)
- Visited marker (3)
- Zoom control (3)

Chapter 1

Getting started with design

Great design rarely happens because of a stroke of genius. It is a creative process that iteratively develops, tests, and refines ideas on how to solve a problem. The process starts with identifying the problem and ends after shipping the app. It roughly consists of two steps (figure 1.1). The first step, finding the right problem to solve, is to explore the problem space. The second step, finding out how to solve the problem right, is to explore the solution space. The reason why both steps are represented by diamonds is that they require divergent thinking to discover possible ideas and convergent thinking to narrow those ideas down to the best idea. The point in between the two diamonds is the definition that frames the design challenge based on a thorough analysis of the problem. One of the greatest mistakes is to omit the left diamond, the problem space, and end up solving the wrong problem.

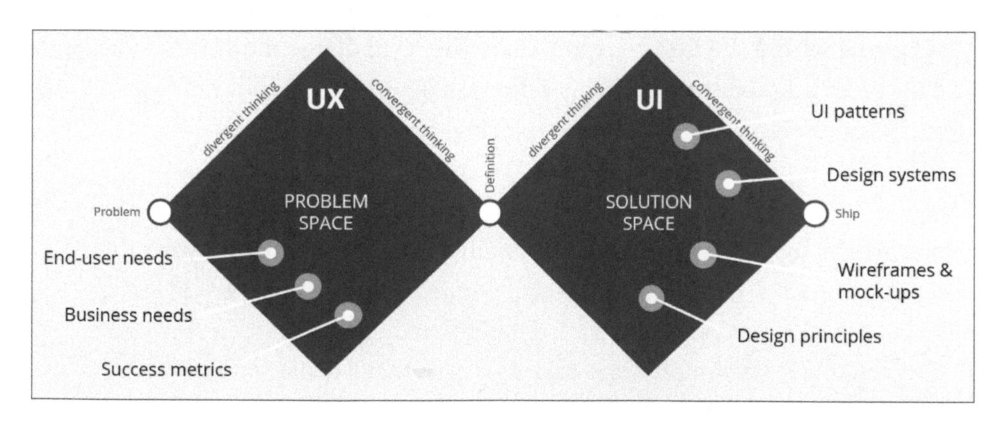

Figure 1.1. Double-diamond design methodology.

This chapter introduces important user interface (UI) design concepts that will help you create more attractive apps. It also shows how thinking through the user experience (UX) will lead to increased usability and ultimately a successful app.

Defining the problem

We often judge apps by the way their UIs look. But that's only half the truth because UIs look good only when they feel good, and they feel good when they are useful. In other words, ugly but useful trumps pretty but pointless. To create something useful, you will first need to explore the problem space (first diamond in figure 1.1), which includes a cumulative understanding of the following:

- **End-user needs:** Who are the users? What are they trying to accomplish? Which problems are they trying to solve? How successful are they in doing that?
- **Business needs:** What are the business's needs? Why are we expending time and money in doing this? What is the purpose of the app?
- **Success metrics:** What does success mean in this instance? How do we define success?

UX means to understand the users and their behaviors, attitudes, emotions, and motivations to use your app. To gain this understanding requires UX research, such as conducting interviews or facilitating workshops. Keep in mind that what the users need and what the business requires can often be diametrically opposed. What users want may not be good for the business, and what the business asks for may be disliked by users. Balancing the two is an aesthetic but important task. Designing a great UI manages to marry both needs into a cohesive whole.

The goal within the first diamond, therefore, is to define the area to focus on and the right problem to solve. This definition is sometimes referred to as UX strategy and serves as the vision for exploring the solution space.

Wireframes and mock-ups

Exploring the solution space (second diamond in figure 1.1) typically involves designing potential UI solutions through wireframes and mock-ups in an iterative manner. It starts with conceptual design, then gradually increases visual fidelity to colors, typography, and dimensions so that the design moves from sketch to photo-realistic and cumulates in ongoing support of and collaboration with front-end development efforts.

Wireframes are sketchlike illustrations of the UI, similar to the images used to introduce the patterns in this book. Their purpose is to communicate ideas and workflows, validate design assumptions, and identify gaps and holes in the requirements. It's faster and more cost effective to make changes to the wireframes than to spend time later to reconfigure the app or even rewrite code. Wireframes are the most important UI design artifacts and serve as the contract between the client and the developer for what needs to be built. A popular wireframing tool is Balsamiq software. It allows you to sketch low-fidelity wireframes using ready-made symbols such as the map controls collection (https://wireframestogo.com/bbd2-Map-Controls), which includes many of the patterns introduced in this book.

Mock-ups are static images of the UI that look close to the real app, featuring many of the final design elements such as brand colors and typography. They aren't functional, but most mock-up software such as Figma allows designers to link individual pages to create interactive workflows that demonstrate how a user might interact with your designs.

Design principles

Design principles are a set of considerations that form the basis of usable UIs. The following list summarizes the multitude of recommendations available in the design space:

- **Guide users to solve a problem:** Make it clear what users should do next.
- **Use familiar and engaging components:** Well-established platform conventions make it easier to learn a new UI.
- **Provide clear and self-descriptive workflows:** Let users control the pace and sequence of interactions.
- **Align the UI with the user's mental model:** Conform to the user's expectations and be consistent.
- **Be tolerant of errors:** UIs should be forgiving and assist users in recovering from errors or prevent errors in the first place.

Nielsen Norman Group hosts the following resources for further reading:

- "10 Usability Heuristics for User Interface Design" are broad rules of thumb for interaction design.

- "5 Principles of Visual Design in UX" helps create more beautiful design and increases usability when applied correctly.

Finally, you don't need to be a graphic designer to create a visually pleasing UI; use a design system as described in the next section.

Design systems

A design system is a set of standards that help guide your design and development. It defines the brand identity, including colors and fonts; provides a ready-to-use component library; and documents design principles and best practices. The reasons you should use a design system are as follows:

- Consistency, both visually and functionally
- Scalability for faster and cheaper design and development
- Ability to focus on larger problems
- Unified language within and across cross-functional teams
- Educational tool and reference

The biggest benefit of a design system, such as Esri's Calcite Design System, is the existence of a component library, which is a set of premade UI elements that have the same look and feel and therefore frees you from dealing with difficult design decisions. Because the components are based on well-established patterns, you don't need to reinvent the wheel to build and maintain your own components. They are available and ready to go.

Calcite Design System includes a UI kit, icons, color schemes, and a web component library with UI elements such as buttons, panels, accordions, alerts, and more. Learn more at developers.arcgis.com/calcite-design-system.

Patterns

UI patterns provide the knowledge about common design problems and how to solve them. This knowledge should inform your design decisions and be applied throughout the whole app design process, from UX research to UI design.

The patterns in this chapter focus on general UX best practices. They explain how to provide clear and digestible entry points into your app and how to guide your users and ease them into workflows:

1. Landing page
2. Task oriented
3. Empty state
4. Focal point

The second set of patterns focuses on searching for locations and data, which are important functions that allow users to engage with your app. These patterns are common and familiar so that people find them easy to use:

5. Search
6. Location finder

You can find additional patterns related to this chapter online. These patterns describe how to improve the UX for first-time users and how to add delight to your app:

7. Onboarding (www.mapuipatterns.com/onboarding)
8. Progress indicator (www.mapuipatterns.com/progress-indicator)
9. Browse geographies (www.mapuipatterns.com/browse-geographies)
10. Spark maps (www.mapuipatterns.com/sparkmaps)

Landing page

Figure 1.2. Landing page pattern.

WHAT

The **landing page** (figure 1.2) is your opportunity to create a great first impression and helps set the tone of the app. It also provides a place for users to input their location so the app can start at an extent that is meaningful and focused.

WHY

For years, builders have chosen to show a map when the app starts. Without further input from the user, the initial map extent must default to the smallest possible scale—for instance, Central Europe for any point that falls within. Typically, this results in an unacceptable UX for two main reasons: first, the user wants to view their own location, such as their state, home, or current location. Second, showing all the data at a small scale is overwhelming, feels crowded, takes longer to render, and, most important, is not relevant.

The landing page is where people start their journey with your app. It's the place where they get introduced to the purpose of your app, and, as in real life, first impressions count. Users make up their mind in a matter of seconds whether it's worth spending more time on this app, or else they abandon it and look elsewhere.

Since the landing page is the first touch point with the app, it's a great opportunity to ask users for their initial input, such as the location of interest, and then use this information to funnel them into a workflow that provides relevant next steps or even answers. Using the landing page to introduce users to the app and gather their location will help optimize the subsequent UX and user interactions with the map.

WHEN

Although landing pages are important entry points for any app, they are most often applied to focused, single-purpose, public-facing apps that are used by a wide variety of people with the goal of finding answers to specific questions. Good examples of sites that facilitate landing pages can be found in the travel, recreation, and real estate industries. The main input is a person's location of interest, such as their travel destination or home. This information can be used to create **task-oriented** workflows that show more results and allow further interactions with that location.

HOW

To create a landing page, add an extra page or full-page overlay before the user reaches the actual map app. This page consists of a few elements and a generous layout with bold, emotional images, appealing typography, and concise copy.

The main measure of success for a landing page is whether a visitor becomes an active user and continues engaging with the app or abandons it. The best way to retain users is to research the target audience, understanding who they are and what they want or need when coming to the app, and offer a clear way to solve this end-user need.

Since the goal of the landing page is to capture and funnel the user into a workflow, it needs to be brief, concise, and attractive and offer fewer distractions. Start by removing anything that clutters the UI and distracts from the main goal, such as links to external sources or global navigation. Instead, provide a clear value proposition and include critical elements such as a headline, tag line, supporting image or illustration, and input to gauge the area of interest. Good input mechanisms to find places of interest are **search** or location finder, and to receive the user's location, use locate me. It's common to place location finder and locate me next to each other in the UI. Another possibility is to provide a series of links to common locations (placemarks). These shortcuts are especially important on mobile devices in which typing search phrases can be cumbersome and annoying.

EXAMPLE

With thousands of miles of trails, the state of Ohio offers nearly endless opportunities to explore the outdoors. The aim of the DE**TOUR** app (figure 1.3) is to prepare outdoor enthusiasts with useful and authoritative information for their next trails adventure. The app starts with a landing page that allows users to search for trails, recreational land, or cities. The landing page also lists shortcuts to popular destinations as secondary input. Both search and placemarks direct the user to the main app that is focused on the selected place. The landing page also offers a list of featured routes to spark interest in exploring something new. Selecting a place or route takes the user to a page that shows more details on the trail and its surroundings.

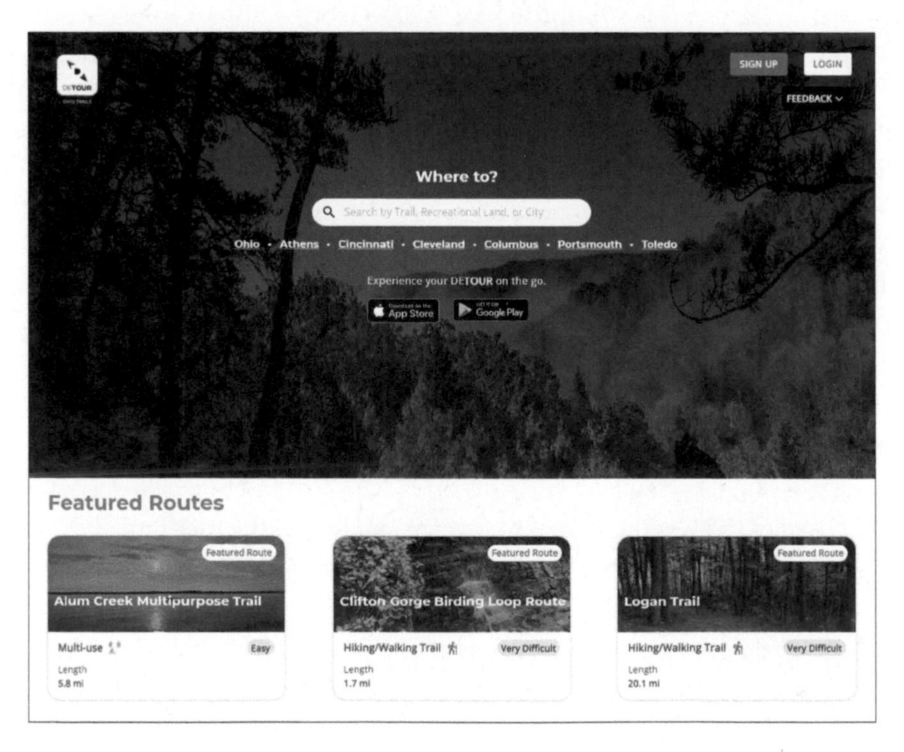

Figure 1.3. Landing page of Ohio's DE**TOUR** trails app.

Task oriented

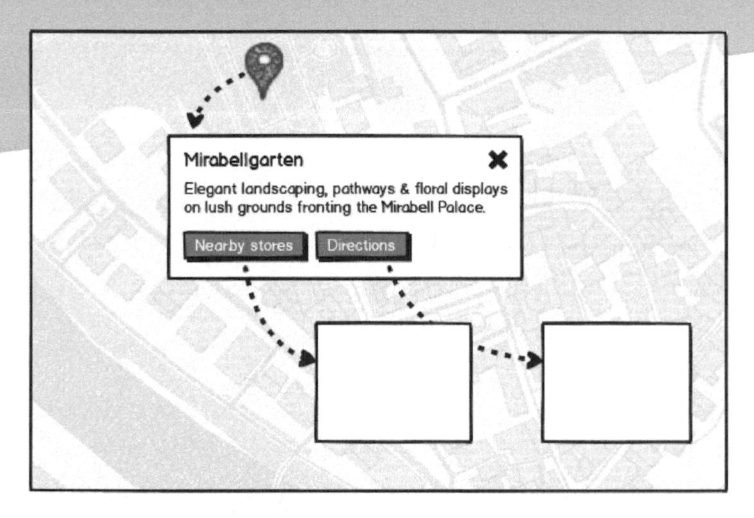

Figure 1.4. Task-oriented pattern.

WHAT

Task-oriented apps (figure 1.4) provide task-oriented workflows that align with user goals. A task-oriented app is sometimes referred to as task focused or workflow driven. Regardless of terminology, a successful experience is built on useful tasks and guiding users to complete them efficiently.

WHY

Task completion and task success are the main metrics for measuring usability and are essential to the success of an app. A single-purpose app that works well is better than an app that tries to solve too many problems at once and is confusing to users. Users get frustrated or even abandon and leave the app if they can't find and complete what they are looking for. As a result, the task completion rate decreases and affects the usability of the app in a negative way.

WHEN

A good indicator that a map app lacks task-oriented design is when it looks and behaves like a toolbox or Swiss army knife. See the kitchen sink pattern, illustrated in chapter 7, "Common Mistakes and How to Avoid Them," for more details. Often, this confusion happens when the app tries to solve too many problems and offers too many features. This lack of focus inevitably waters down the purpose of the app, and as a result, users are confused where to start, what to do, and how to accomplish their tasks.

HOW

To be task oriented, funnel users into a clear workflow that assists them in successfully completing their task. Avoid scattering tools and functionality across the UI but chain them into a single task-oriented workflow instead. This workflow can be triggered in different ways.

One approach is to provide a single call to action (CTA), such as a button that opens a UI container (for instance, a panel or modal window), with the goal to focus the user on completing the task. A modal window is a child window that opens in front of the main window while blocking the rest of the app. Another way is through explicit action buttons inside an **info pop-up**. This approach is especially effective because you can deduce the intent of the user from their clicking a feature and provide workflows that match this intent and context. Having the workflow readily available in a final panel avoids the need for unnecessary interactions to reach it. **Partial map** layouts work best in this scenario. Another good practice is to embed tools that interact with the map such as **feature selection** right in the workflow so users don't need to look elsewhere to "click the right button to turn the right knob."

Some common task-oriented use cases are discussed in chapter 6, "Building Single-Purpose Apps."

EXAMPLE

The Riverside County Office of Education (RCOE) hosts a custom app (figure 1.5) that allows employees to evaluate school transfer requests by comparing the distances from applicants' home locations (such as Dad's, Mom's, Grand-parents') to the home school and the transfer school. The task-oriented nature of the app provides steps to first select the home school (abbreviated by an *H*) and transfer school (*T*), followed by entering one or more address locations of the applicant (indicated by letters). For each location, the app then calculates the distance to the home and transfer school and shows their respective routes on the map. Each route is displayed in a different color that also correlates to the location card in the left panel. This layout, in which the main workflow is prominently represented in a panel and the map appears next to it, is called **partial map**. Users can interact with the location cards—for instance, clicking on a location card on the left will zoom the map to the extent of its related routes. An interesting usability detail is that the distance to the target school does not show the actual distance in miles but instead shows the difference between home school and target school—such that a negative number indicates that the distance to the target school is less than the distance to the home school. This seemingly minor detail helps assessors make easier and faster decisions.

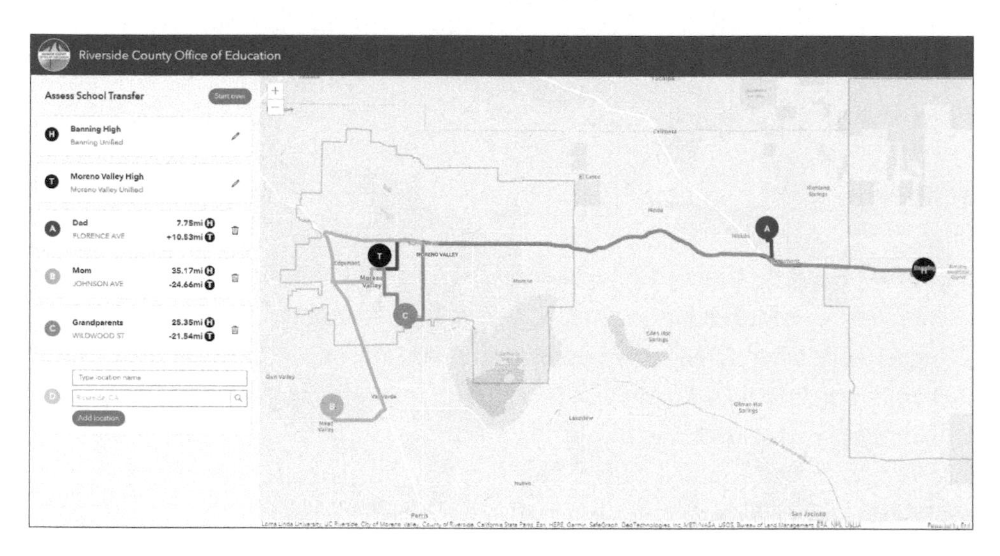

Figure 1.5. App to assess school transfer requests by comparing distances from applicants' homes.

Empty state

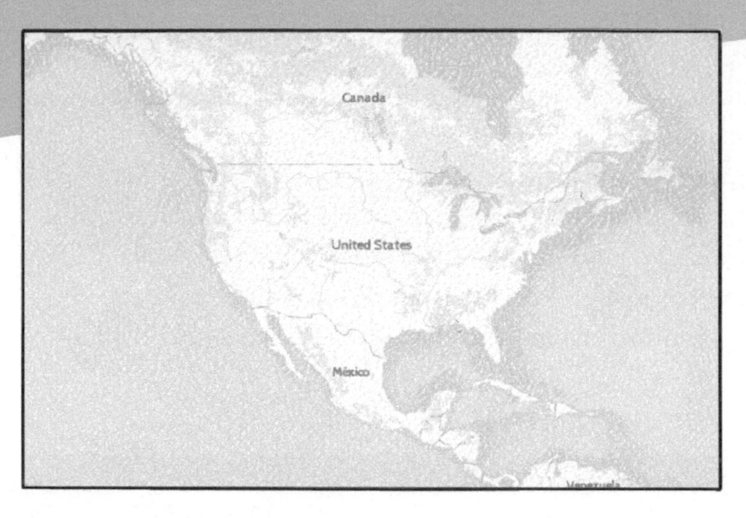

Figure 1.6. Empty state pattern.

WHAT

The **empty state** (figure 1.6), sometimes also called blank slate, is one of the most overlooked aspects in UX design. It refers to moments during the experience when screens, panels, or containers are empty. Leaving something blank or empty does not make a good first impression and leaves the user wondering what to do next.

WHY

The most common reasons for empty states are first use (user hasn't started the task to generate data yet), no results or no data (search or filter returned zero results), or the user has finished their task or wants to reset the app to its original state, such as clicking the **home button**.

In the context of a map app, it's easy to get the impression that a map is not empty because it shows colorful content when the opposite is true. The starting extent of the whole country or continent is rarely relevant to a user. It often shows too little (empty) or too much (overload) information and is as meaningful as not showing a map at all. Therefore, considering what the empty state should be is crucial for building an effective map app because it serves as an important (first) touch point between the user and the interface.

WHEN

A common mistake in map apps is to default to an extent that is at the smallest possible scale, such as showing the whole country or state. In many cases, this magnitude turns out to be meaningless to the user whose need is to discover or research details about their area of interest. Additionally, the map will most likely display unfamiliar data or even be empty, with little guidance on the next step. As a direct result, the user may feel intimidated or lost and, in some cases, even think that the app is broken.

Builders need to consider the empty state for any type of app but it's especially important during initial use. Other common use cases include potentially empty data displays such as lists or search results.

HOW

The main goal of dealing with an empty state is either to fill it with useful, helpful, and informative content or avoid empty spaces in the first place. Filling empty space is an opportunity to communicate. It provides you with the space to teach users how to use a new feature, guide them in getting started with a task, or tell them why something is empty. The latter is an opportunity to show the character of your app by adding creative placeholders or a funny image that delights the user. Resist the urge to "reserve space" just because it might be needed later.

A map can feel intimidating and overwhelming, so it's important to show empathy with your users, understand the purpose of their visit, and think about what can be done to encourage them to interact with the map. Simplify the map as much as possible, provide clear labels, add **markers** or a **focal point**, and consider other solution strategies as outlined in chapter 4, "Dealing with Complex Data."

One component that does not have a case of empty state is the **info pop-up**. If the user clicks in the map without touching a feature, there is no reason to report the absence of data. This type of result is annoying, requires unnecessary work to close again, and should be avoided. In cases in which the app can determine the user's intent for clicking in the map or if any inaccuracies may have led to the empty result set, the info pop-up could instead provide a solution to help the user move on from the potential input error.

Finally, another approach to avoid the empty state fallacy is to implement a **landing page** that prompts users to specify a location up front, which allows the app to start at an extent that is meaningful and less intimidating.

EXAMPLE

Figure 1.7 shows an example of a **store locator** with all the interface elements you might expect: search input, filters, search results, and a map. Even though the designer thought about the empty state and tried to accommodate by using the blank space of the search results and adding an informational message, "Enter your search above to see results," it still feels like wasted space. The filter isn't useful because nothing has been searched for yet, and the map doesn't invite the user to use it as an alternative input. All in all, the app looks and feels bare.

For the redesign (figure 1.8), all the elements that haven't been used yet are removed, which helped provide space to add a tag line, "Find a store near you!" and potentially links to featured stores or special offers. The redesigned map now highlights states that have stores and invites users to click in the state of interest as an alternative means for input. In summary, thinking through the empty state helps optimize the UI and improve the UX.

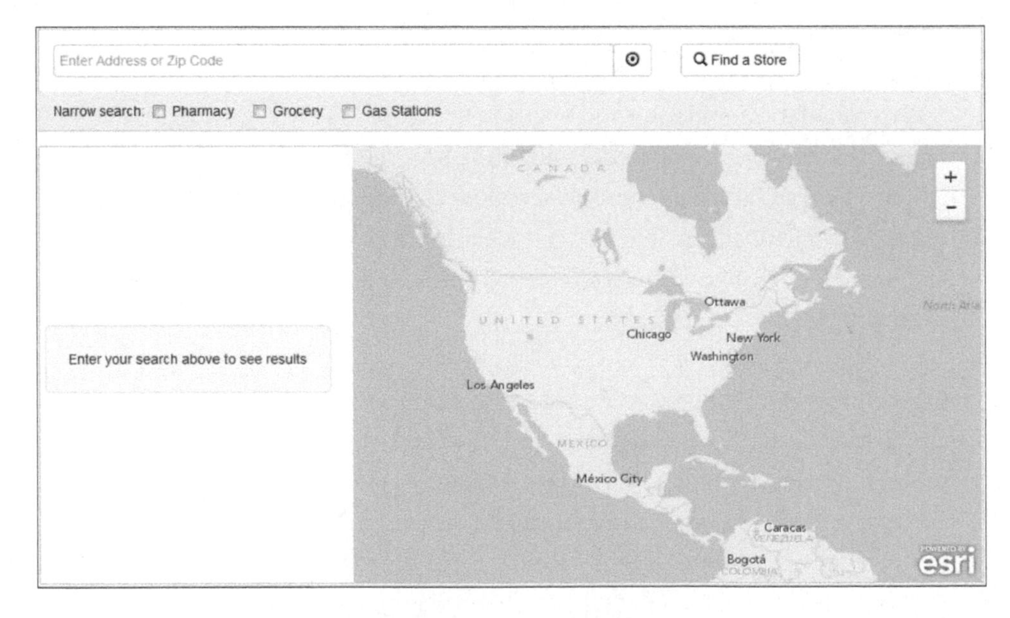

Figure 1.7. Empty state of a store locator app before redesign.

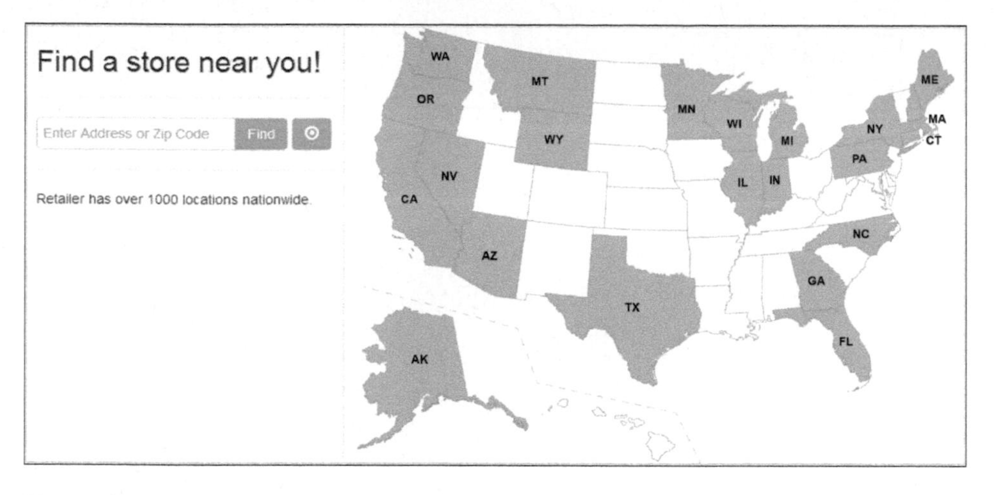

Figure 1.8. Empty state after the redesign.

Focal point

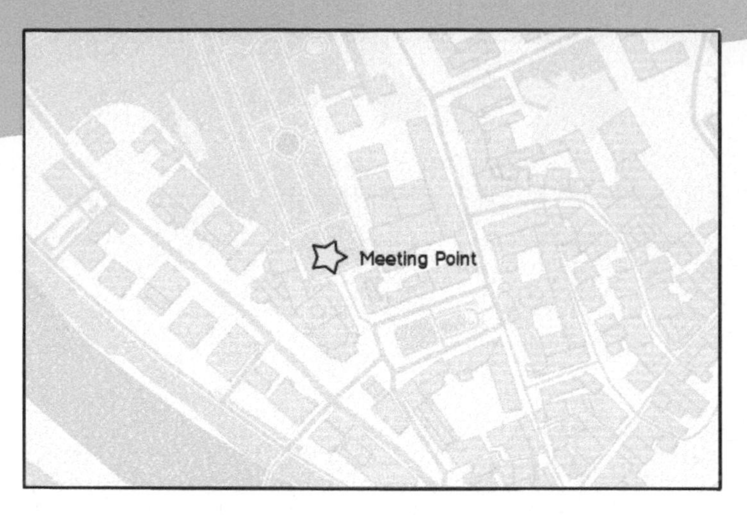

Figure 1.9. Focal point pattern.

WHAT

The **focal point** pattern (figure 1.9) highlights the center of interest or activity and assists the user in finding the important piece of information quickly.

WHY

Harmony in design can be attributed to the gestalt theory, a set of design principles that describes how humans perceive groups of objects. At its core, gestalt theory states that the whole is more than the sum of its parts. If applied correctly, then we as humans will see a map of a city and not its individual parts, such as buildings or streets, and that's desired because it would otherwise lead to cognitive overload. On the flip side, there is nothing natural, special, or relevant in a map that stands out and draws our attention unless it is something that's being highlighted.

Adding a focal point to highlight a feature on the map takes out the guessing game and relieves its users of searching for the needle in the haystack. It is an element that adds emphasis to a point of interest or activity with the intent of drawing the viewer's attention.

WHEN

Probably the most common use of a focal point is adding a **marker** to the map to show the location of the place after a **search**. Another common use is for navigation apps to show two focal points that indicate the start and end points of the route. Emergency apps or media outlets use a focal point to highlight the epicenter of an event such as a plane crash or an earthquake.

HOW

The goal of the focal point pattern is to visually emphasize a feature on the map to draw the user's attention. This feature is typically a point (city, landmark), line (river, highway), or polygon (state, country) that is well known and easily recognizable. It might also be the result of a prior task the user has performed, such as **search** or **feature selection**.

There are three types of focal points:

- **Markers** added to the map that stand out through their position (map is centered on the marker), their size (bigger than the surrounding features), their shape (using an unusual shape such as a star), or their color (a high-contrast shape that is clearly different from its background)
- Features of the map that have effects applied to them, such as drop shadows, glowing edges, or animation effects
- Masks that hide irrelevant features

When working in real time or in near-real-time environments, the focal point is usually an animated marker such as a pulsating dot or moving shape. The animation indicates that the location is updated frequently and can change anytime.

EXAMPLE

Geo-Referenced Infrastructure and Demographic Data for Development (GRID3) is a nonprofit program funded by Columbia University that helps countries create and use geospatial data on population, settlements, infrastructure, and boundaries. GRID3's Data Explorer tool (figure 1.10) makes it easier to choose the most appropriate geospatial dataset for practical applications in sub-Saharan Africa. When a user selects a country, such as Zimbabwe, they can explore all available datasets sorted by data theme. The user then can compare metadata across data sources, visualize datasets through interactive web maps, and download datasets from the distributor.

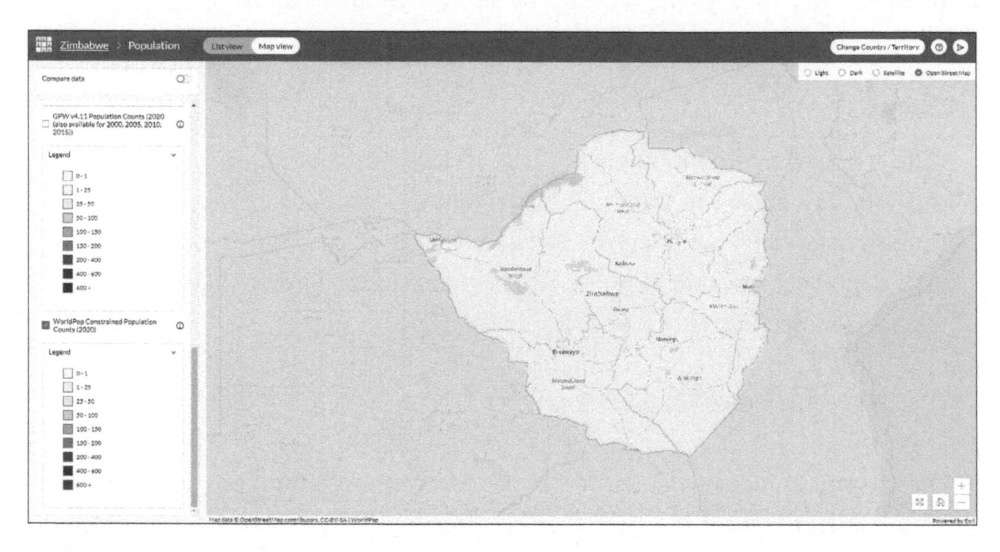

Figure 1.10. GRID3's Data Explorer tool allows users to find the best available dataset.

Source: GRID3.

Search

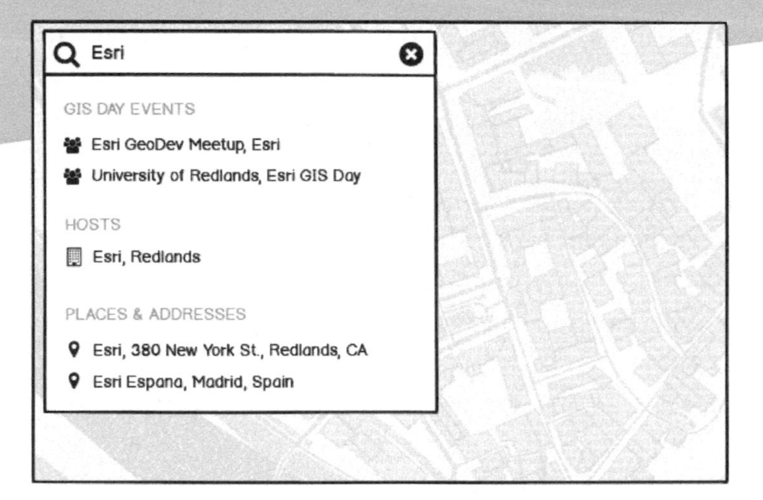

Figure 1.11. Search pattern.

WHAT

The **search** pattern (figure 1.11) uses a single-line input field for users to search for a location. It is the main mechanism for finding a place or an entity that has a location associated with it. Users expect to receive the correct result, regardless of what they type and search for.

WHY

The search pattern is often the first step in a user's journey and may lead to extended workflows. Search is probably also the worst usability problem to solve—it's complex and difficult. What should you type or enter? You're on your cell phone or in front of a kiosk, and typing is frustrating. Why aren't there any results? Why are there so many? The thing you're searching for doesn't show up so it must not exist.

That's why universal search engines have become the go-to address to find information. People have grown accustomed to using a single-line input field to retrieve aggregated results from different data sources, regardless of the spelling, length, form, or language of their search term. But not everyone has the resources to build comprehensive search functionality on that scale.

The key to a great search experience is to present a single set of results that is free of any duplicates but gives the user easy access to the diverse information contained by the different sources of available data.

WHEN

Search is the logical first step in a user's journey. It's faster to search for a place of interest or an address than zooming and panning the map multiple times (see **location finder** in the next section). It's also easier to search for such things as events, depots, replacement parts, hospitals, or incidents than clicking each individual marker on the map to find out what it is. Combining the search for places and things into one unified search helps users find what they're looking for in one action.

HOW

The search function is complex, but the two most important keys to success are as follows:

- **It's fast:** Users want to see immediate results with minimal response times. Your hardware and algorithms need to be optimized to meet this expectation.
- **It's federated:** As data becomes more distributed across providers and services, people are used to entering a keyword or phrase without having to know where the results come from—and they shouldn't need to know. Don't make them select a data source from a list before they can type their keywords.

The search interface is usually a single input field, sometimes with a leading magnifying glass icon to indicate its purpose. The search may open a list of actions such as **locate me** and suggestions of previous searches or personalized places, such as your home or work. As the user starts typing a search term, the search engine queries content from different sources and aggregates and disambiguates the results into a single set of search results. Providing an autocomplete feature is recommended, in which the search engine predicts the user's query and provides suggestions as the user types. This feedback mechanism allows users to detect mistakes such as typos before they submit a long search term that returns zero results. On the flip side, the constant queries put a strain on the system architecture, so it's advisable to add a waiting period of approximately 30 milliseconds between keystrokes before the search engine fires the query.

Once the search results come back, it's important to disambiguate them in a meaningful and logical way that includes removing potential duplicates, weighting results by relevance or importance, indicating the source or nature of the result, and completely removing excess results. Structure the results by grouping them into categories and adding as much information as possible. Consider highlighting the characters that match the search term.

Make the search result actionable and indicate this to the user by changing the mouse cursor to a pointer. Upon the user clicking the result, zoom the map to full extent. If the search result is a point, consider replacing the point with a temporary **marker** and possibly opening the associated **info pop-up**. If the result is a line or polygon, highlight or emphasize it in some way.

Not every result has a spatial component (geometry) associated with it. Searching for a person in a directory will pull up the employee, but another query is needed to relate this person to their office, which can be located on the map.

EXAMPLE

As a government entity, GIS and Data Services, City of Lubbock, Texas, needs to cater to a user base with a wide range of needs. There are Realtors interested in zoning or evacuation zones, mortgage underwriters interested in hazard zones or road ownership, and citizens who want to look up miscellaneous information. Instead of building multiple single-purpose apps, the City of Lubbock decided to build an app (figure 1.12) based on the assumption that people are familiar with searching, so it is easy for them to start typing whatever they need to find. The autocomplete function then returns results grouped by interest (parcels, addresses, roads). Once a result is clicked inside one of these categories, the app funnels users to the appropriate workflow. For instance, if the user selects an address, the workflow will show details of the property with the option to run an analytic tool to show more information about the property's surroundings. The app uses **full map** layout to maximize the map space, whereas all the other map tools are tucked away in the navigation panel. Clicking the menu button opens this panel in which users can toggle layers on and off or change the **basemap** layer.

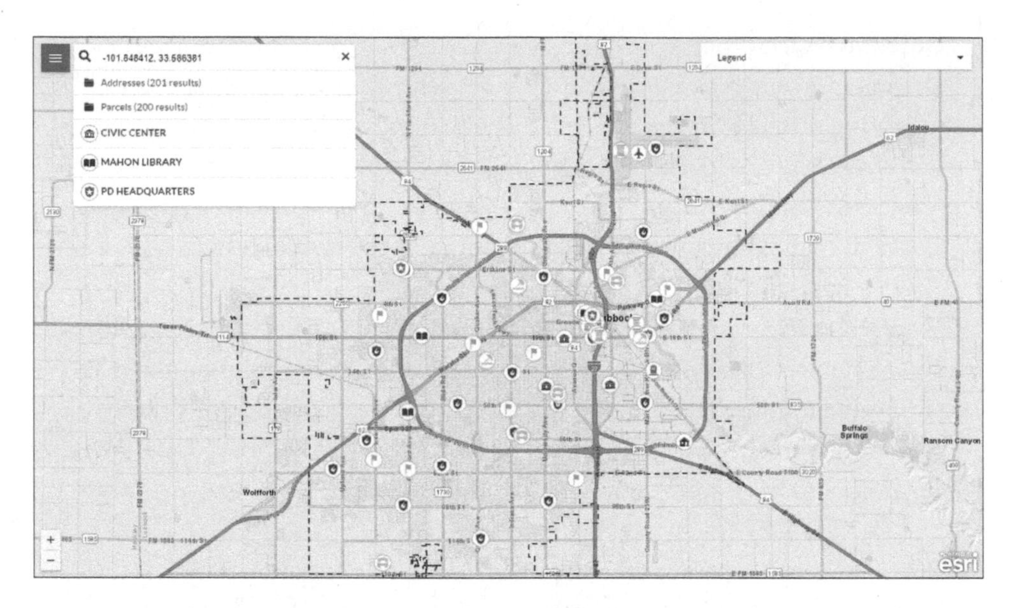

Figure 1.12. City of Lubbock offers users an app to explore data in and around the city.

Source: GIS and Data Services, City of Lubbock, Texas.

Location finder

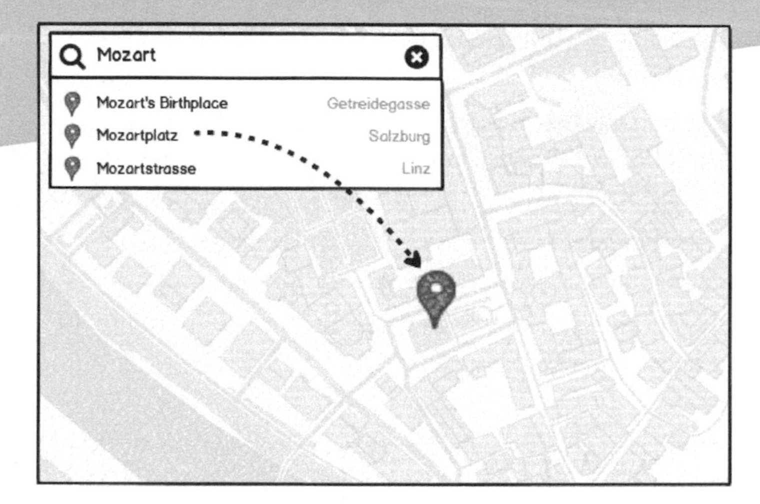

Figure 1.13. Location finder pattern.

WHAT

The location finder pattern (figure 1.13) helps find a place, point of interest, or address and shows it on a map.

WHY

Location, location, location. What has been the number one rule in real estate for many years is also true for map apps. Showing a map is valuable but showing a map that is centered on the user's area of interest is more valuable. That's why the process of geolocation to identify the real-world geographic location of an object is the most important piece of any map app. It's typically the first task a user is looking for before they start any sort of spatial analysis.

WHEN

It is good practice for any map app to offer a location finder that zooms the map quickly and easily to an extent that matters to the user. Finding a place, point of interest, or address is a meaningful first step in the user's journey. Once the location is identified, the map can zoom to a meaningful area and scale. The user can then go about a task, such as exploring or analyzing the area, inspecting points of interest, or updating map data.

HOW

Provide an input field that allows users to type their search. Often, the location finder has a leading magnifying glass adornment. As the user starts typing, the autocomplete functionality is a good way to show potential locator results and prevent typos. After the user selects the desired result, the results list collapses and the search term is replaced by the selected result label. A reset button is added toward the end of the finder input to clear the finder. Zoom the map so it is centered on the coordinates of the location. Choose an extent that provides enough context to validate that the location is the desired result. Add a **marker** or outline the area with a polygon. Opening an **info pop-up** to invoke further actions is a common practice, as well.

The search results should list the most common location associated with an unqualified name first—for instance, if *Rome* or *Paris* is specified, Rome, Italy, or Paris, France, will be listed rather than a less commonly known place.

If the search returns exactly one result, the application may treat this finding as the desired result and automatically center on it. Another school of thought is that explicit user action is preferred, and the application should wait for the user to click the result before zooming in.

After a new search is invoked, any existing marker on the map will be removed and placed in the new location. Once the user starts zooming and panning the map, the marker should remain on the map until the selected result gets cleared by clicking the reset button in the location finder.

EXAMPLE

Arizona Public Service (APS) generates clean, reliable, and affordable energy for 2.7 million Arizonans. Its Outage Map app (figure 1.14) allows customers to find electrical outages that impact their neighborhood. In most cases, users are mainly interested in seeing outages around a particular address, and that's why the app, which wants the **location finder** to be the first input element in the user's natural reading order, placed it in the top left corner of the map. After performing an address search, the app adds a **marker**, zooms to a medium scale, and centers the map at that location. The reason the map zooms to a medium scale is so that users can visually identify nearby outages without having to zoom and pan. They can then click on an outage to open an **info pop-up** and receive more information. The outage affecting the customer's address at West Indian School Road, for instance, was caused by a planned outage. The pop-up also informs affected residents that crews are already on-site to proactively upgrade equipment to help ensure reliable power is available, with an estimated restoration time of 11:20 a.m.

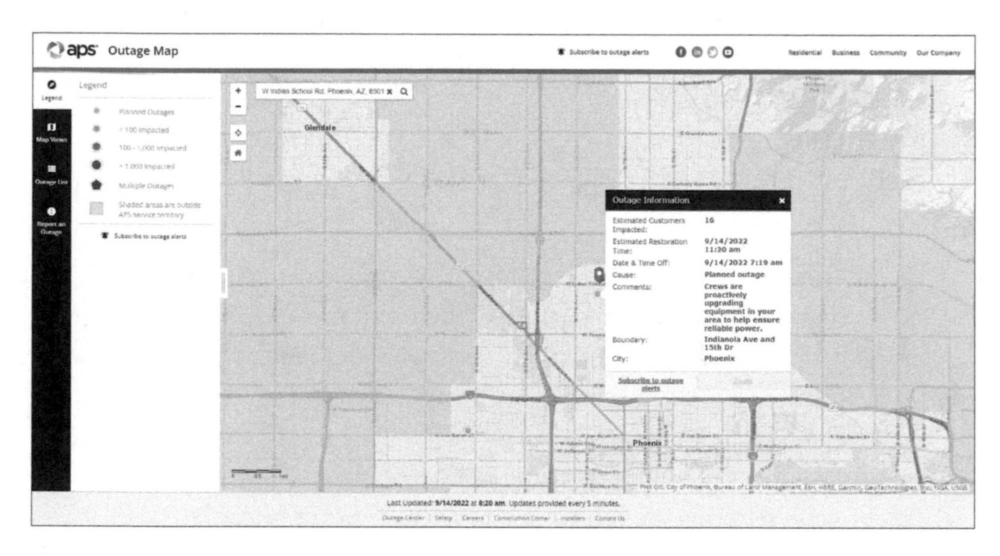

Figure 1.14. APS Outage Map shows electrical outages around an address.

References

"5 Principles of Visual Design in UX," www.youtube.com/watch?v=hO22jAehY1E.

"10 Usability Heuristics for User Interface Design," www.nngroup.com/articles/ten-usability-heuristics.

Balsamiq, https://balsamiq.com.

Balsamiq Map Controls, https://wireframestogo.com/bbd2-Map-Controls.

Figma, www.figma.com.

Search, https://developers.arcgis.com/javascript/latest/api-reference/esri-widgets-Search.html.

Chapter 2
Selecting the right layout

Layout deals with the arrangement of visual elements in the UI and requires a good understanding of their purpose and usage. It is typically the first step during the conceptual design phase of an app and can easily feel overwhelming when thinking about the wide array of UI components and how they can be laid out on a screen. This chapter describes the most prominent layout patterns and offers a way to make a well-informed decision on selecting the correct layout for your next project.

Knowing your personas

The first, and possibly most important, step in deciding the correct layout is understanding your target audience. This is done by defining a so-called persona, an archetype that describes a group of users with common goals and abilities. Ideally, this persona is based on user research and insights collected from interviews or surveys.

The following section describes five high-level personas of commonly observed user types. They are not meant to replace your own research and are by no means discrete or complete but provide a rough guide to deduce the right layout based on the applicable app type.

The five personas and the primary goals associated with each are as follows:

- **Expert or subject matter expert (SME):** Needs detailed and in-depth insights
- **Analyst:** Wants to see the complete picture and needs comprehensive analytic tools

- **Executive:** Relies on strategic performance indicators to make decisions
- **Public user:** Looks for relevant and specific answers to their current task
- **Mobile user:** Looks for precise directions and information in real time

As shown in figure 2.1, the abilities of these user types differ from being focused with plenty of time ("Focus") to being distracted and in a hurry ("Attention"). This characterization is based on the nature of their tasks but can also be attributed to the situations these users face. A similar spectrum applies to their area of interest ("Scale"), which ranges from broad interest in several fields for the SME to being focused on a specific area, such as a street address for a mobile user. Overall, it can also be assumed that users with personas on the left side of the spectrum, with high attention and wide focus, have higher proficiency in navigating and interacting with maps ("GIS literacy") than users with personas on the right side, with low attention and narrow focus.

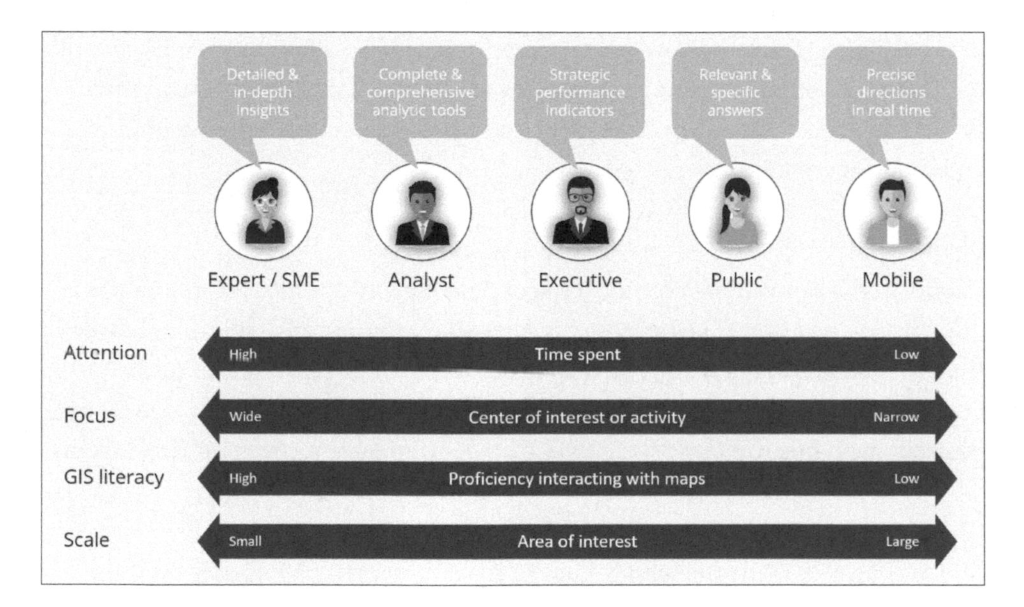

Figure 2.1. Characteristics of the five most common personas.

Types of apps

After determining the persona, the next step is to define the types of apps these people use (figure 2.2): enterprise, analysis, dashboard, single purpose, and mobile. This classification is done according to the tasks these users perform, their abilities, and the situation.

Enterprise

The expert user spends most of their time using the system and its functions. They need maximum flexibility and advanced functionality in the form of an enterprise app. The perfect layout is a **partial map** layout that provides **task-oriented** content side by side with a map. It's okay for the map to show an advanced level of detail.

Analysis

The analyst user is methodical about their tasks, can perform complex queries, and needs tools to perform analysis. Like the expert user, the analyst benefits from a **partial map** layout, but because of their high proficiency and analytic needs, a **full map** layout may also be appropriate.

Dashboard

The executive user relies on analysts to prepare tactical data as input for making decisions. The app type is usually a dashboard that provides at-a-glance insights into business intelligence, key performance indicators (KPIs), and trends, as well as other types of information. Interactions are kept to a minimum. Data visualization elements such as KPIs, charts, lists, and maps support the briefing objective and therefore require a **reference map** layout.

Single purpose

Public user is a placeholder for any type of persona with an interest in public information and data. They are usually focused on finding an applicable answer to the problem at hand, and thus the app needs to be single purpose and to the point. Depending on the actual persona and particular use cases, a variety of layouts can be used, including **full map**, **partial map**, **embedded map**, or even **no map** layouts.

Mobile

The mobile user is anyone who is "on the go," not necessarily because they are using a mobile device but because they rely on accurate and up-to-date information while in a rush. As outlined in more detail in chapter 5, "Designing for Mobile Devices," mobile design requires more than a single layout—it's a series of screens, workflows, and patterns.

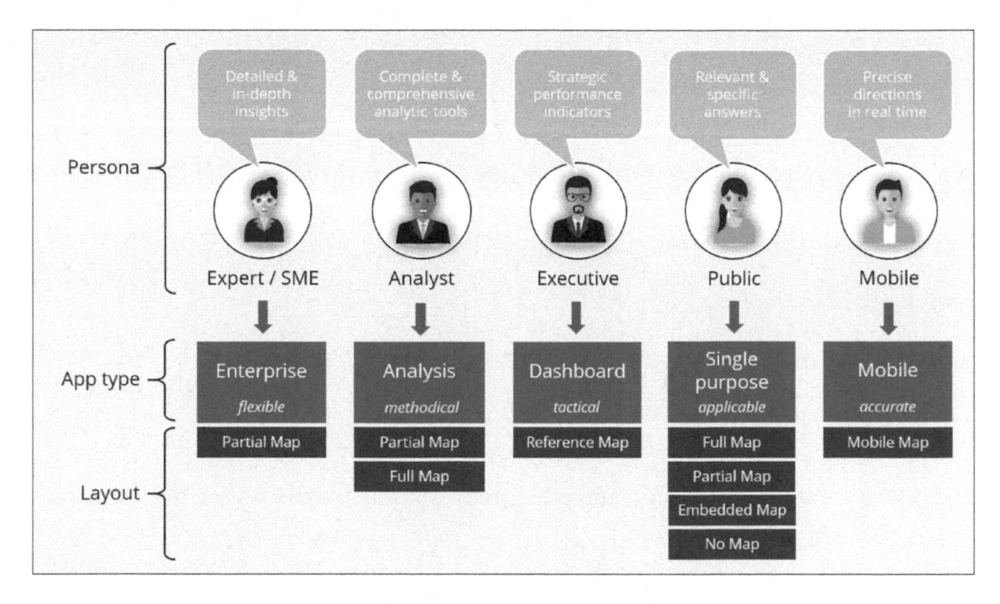

Figure 2.2. Selecting the correct layout based on target persona.

Patterns

Whether you design a custom app from scratch or build an app using an app builder tool, deciding on the appropriate layout can be a daunting first task. Figure 2.3 compares the different layout patterns with their pros and cons. Understanding the suggested level of detail, level of control, number of tools, interactions, and attention to cartography will help you make better choices for the right interface components to use.

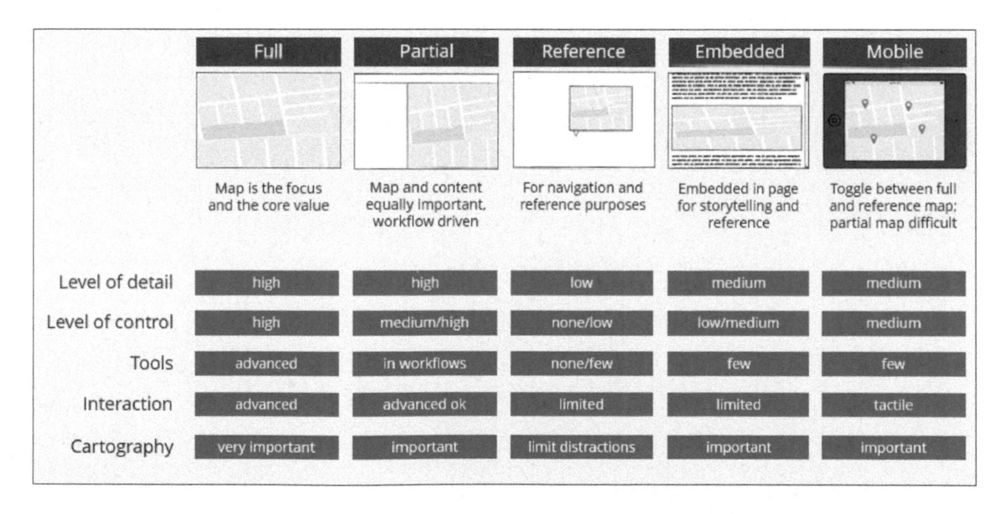

Figure 2.3. Comparison of layout patterns.

The main layout patterns described in this chapter are as follows:

1. Full map
2. Partial map
3. Reference map
4. Embedded map
5. No map
6. Focus on the subject

Some layouts are less common and are used in specialized places, such as airports or shopping malls:

7. Kiosk map (www.mapuipatterns.com/kiosk-map)

Full map

Figure 2.4. Full map pattern.

WHAT

In the **full map** layout (figure 2.4), the map is the focus of the app and large enough to fill the entire screen. The core value of the app is spatial information.

WHY

Spatial analysis allows people to get insights and discover relationships in geospatial data. The way this rich data is visualized directly impacts how users can solve complex location-based problems. The map view is the core value of the app. Providing a larger and more complete view provides analysts and GIS experts with the geographic context necessary. These users spend long periods of time interacting with the data and need to see as much of the map as possible. The full map layout helps maximize the level of detail and minimize the number of distractions from other components.

WHEN

Use full map layout when the value to the user is the map itself, as in performing complex analyses or viewing assets in relation to their current location.

Taking over most of the screen real estate helps focus on the display and analysis of spatial data in an interactive environment.

There are circumstances when full map layout, which often includes other map tools that overlap the map, is not the best choice. The situational awareness, safety, or security industries, for instance, always require their operators to see the complete picture. In these cases, it is not acceptable to cover potentially important data with floating panels, and a **partial map** layout is the better choice.

HOW

Provide an interactive map that takes up all or most of the screen canvas and operates as a single-page app. Place navigation and interactive tools on top of the map, usually in the corners. Arrange these tools in logical groups and position them in a balanced fashion. Limit the number of tools that float on the map to avoid obstructing important content and maximize the visible map content.

Adding a title to the map helps people understand its purpose. This can be accomplished by either adding a separate header section (preferred) or floating text on the map itself. In the latter case, consider adding a drop-shadow effect to distinguish it from the underlying map layers.

Pay extra attention to the cartography, and don't settle for default symbology. Correct and precise data display in full map layouts is paramount to the success of the app. Scale-dependent symbology and labeling helps provide the right information at the right scale.

It's safe to assume that expert users often have higher GIS proficiency. Apps geared to that target audience can therefore apply more advanced mapping capabilities and user interactions.

EXAMPLE

The National Agricultural Statistics Service of the US Department of Agriculture (USDA) uses a **full map** layout (figure 2.5) to highlight its crop data. Instead of having floating panels on the map, the tools are in the banner area, where users can easily access them. The tools are arranged by anticipated sequence of use. The first step is to find a particular area (**location finder**) and then filter the data using either a **spatial filter** (select area and define area) or an **attribute filter** (filter crops). Finally, users want to run the analysis and export data. The only tools that remain on the map canvas are common features that control the appearance of the map itself such as **zoom control**, **home button**, **timeline slider**, **basemap gallery**, and **layer list**.

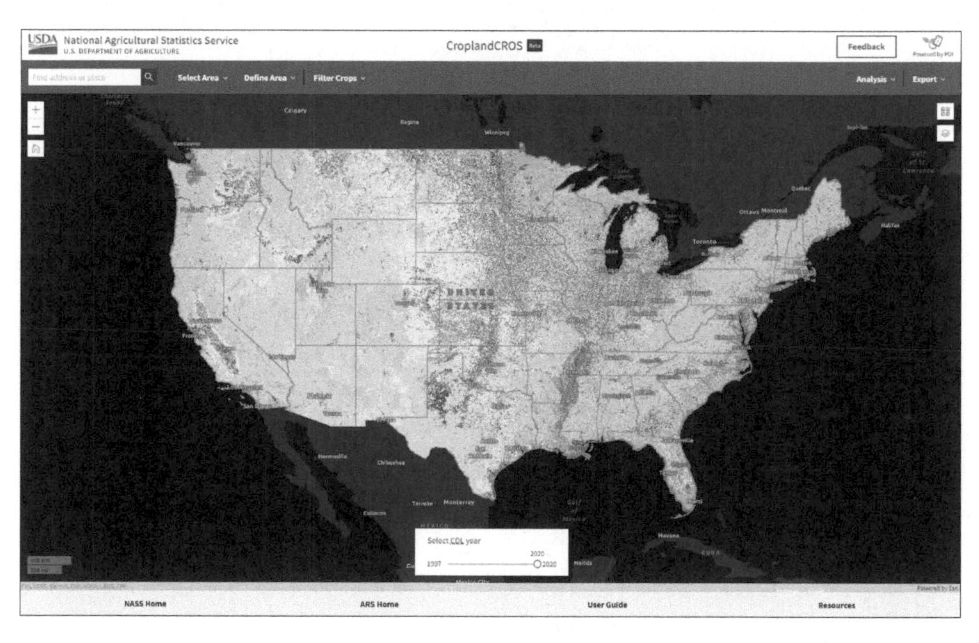

Figure 2.5. USDA CroplandCROS uses a full map layout to show as much of the data as possible. Source: US Department of Agriculture.

Partial map

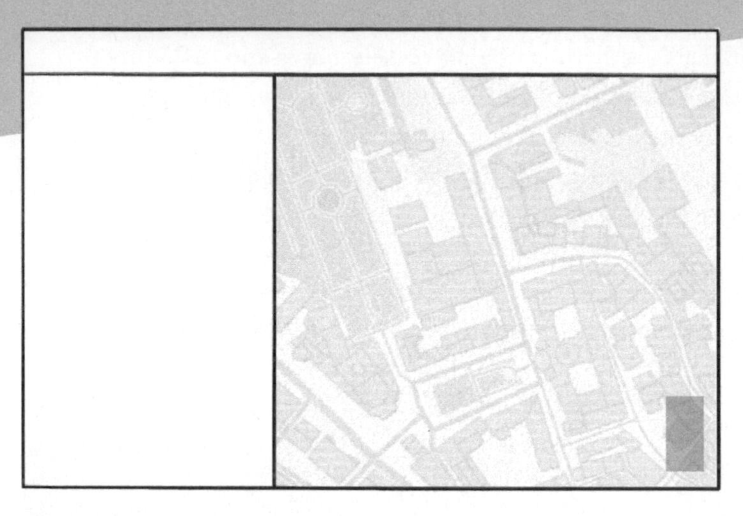

Figure 2.6. Partial map pattern.

WHAT

In the **partial map** layout (figure 2.6), panels for workflow or content are side by side with the map. Map and content are equally important, and the app is often workflow driven.

WHY

As outlined in the **task-oriented** pattern in chapter 1, "Getting Started with Design," having clear and well-defined workflows is important for aligning content and features with the goals of users. Having a panel for content and workflows side by side with the map helps users focus on their tasks and avoids obstructing the map. The map itself often supports the workflow in showing information that is needed for making decisions or selecting and modifying features as part of the task. Therefore, the map and the content are equally important: they support and complement each other.

WHEN

The partial map layout is common in enterprise apps in which business workflows are important and users are GIS savvy and well trained. Use it also when you need to explore or modify data as part of the workflow.

Partial map is also a good layout choice for apps with advanced interactions between the content and the map—for instance, when the map drives the navigation of the app while simultaneously letting the user discover additional information and interact with the data.

HOW

Build the app so that the map appears side by side with the content or workflows. In left-to-right languages, the content panel is typically on the left of the map if it holds workflows but can also be on the right for pure data display. Some apps may place content below the map. The best placement should be decided based on the relative importance to the map. Often, both sides are connected so that interactions on one side trigger the appropriate action or update on the other side, and vice versa. In fact, these interactions are so widely used that they are described in detail within their own patterns, location list and extent-driven content (see chapter 6, "Building Single-Purpose Apps"). Enterprise apps often also add a banner area on top of the content and map. A banner is a good way to include branding, search, attribute filters, and other tools that are not part of the workflow itself.

The real power of partial map layout is the ability to embed map tools into the workflow so that they are available as part of the logical flow and users don't need to look for them in different places. These tools can include sketching a new feature or selecting an existing feature to change its attributes.

Pay special attention to cartography because this type of map is still complex, with a plethora of layers and interactions.

EXAMPLE

The Boston Water and Sewer Commission has developed a citywide model that was used to simulate a variety of wet-weather events and flooding scenarios that could occur in Boston, Massachusetts. The result of this project may help inform the city's efforts to prepare for the effects of climate change, sea level rise, and more intense storms. The Inundation Model Viewer app (figure 2.7) uses the partial map layout, in which the main workflow is inside a panel at the bottom of the screen. The user can follow steps to select from different weather events, adjust the amount of rainfall, and modify the storm surge. Tweaking these parameters updates the map and the impact forecast metrics shown in the last box, on the right.

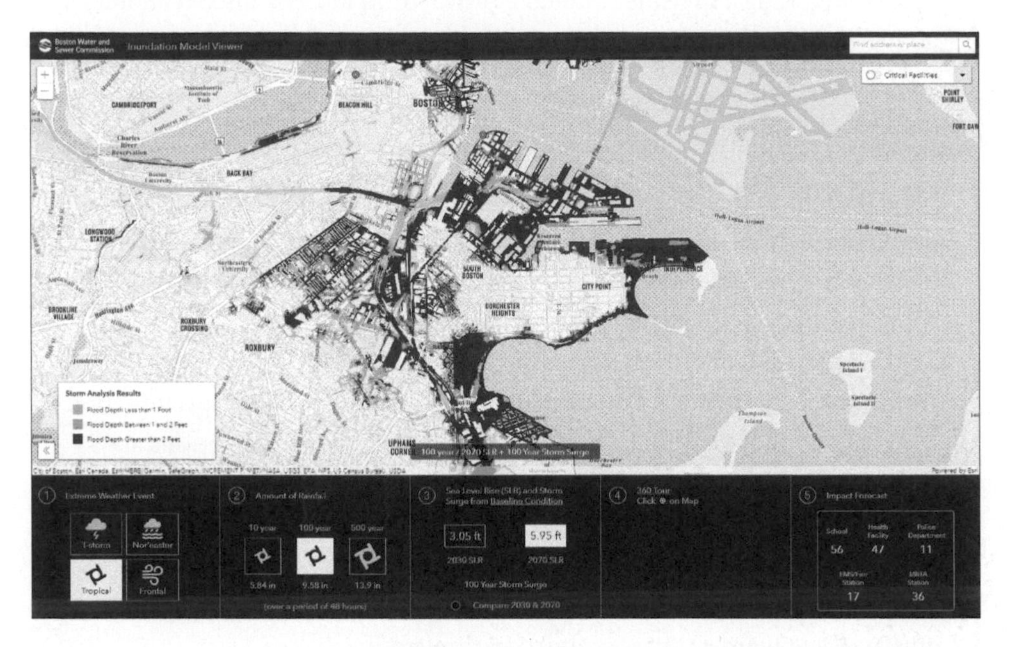

Figure 2.7. The Inundation Model Viewer app lets users select their forecast metrics.

Reference map

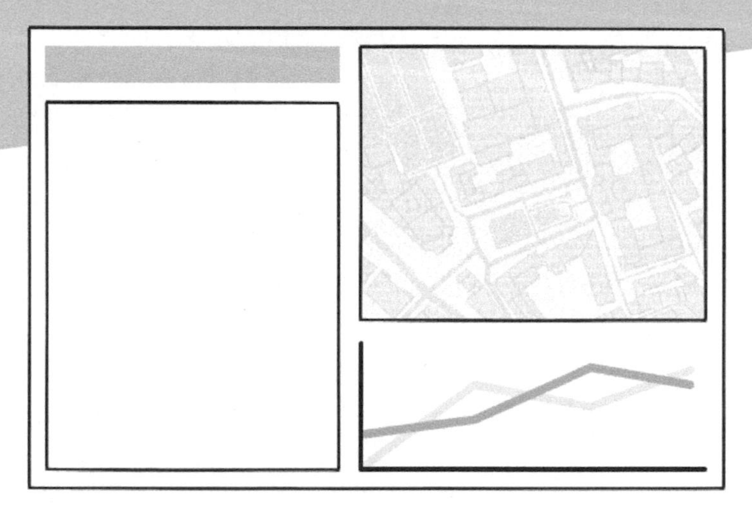

Figure 2.8. Reference map pattern.

WHAT

In the reference map pattern (figure 2.8), a small map is generally placed next to the other layout components. The map serves as an additional or auxiliary view and is mainly used for contextual purposes.

WHY

Geography is important for understanding everyday things. How is an object spatially related to another object or landmarks such as roads or highways? Seeing this location information is important for making rapid decisions, and, as we all know, a picture is worth a thousand words. In most cases, users don't want a fancy map with a lot of tools. All that is needed is a small map that highlights the object in context with its geographic location. Full map or partial map layouts are simply too much.

WHEN

Use when the map adds benefit to the understanding of information shown elsewhere or provides geographic context. Good examples of the reference map layout are dashboards that contain various data visualizations such as

KPIs, lists, tables, charts, and descriptive text. The map then becomes another visualization component that adds value to the data orchestration and is a key element for decision-making. The map itself becomes secondary and less dynamic in nature.

HOW

The purpose of the map in a reference map layout is to display relevant content auxiliary to other information. This map is usually lightweight with limited interactions and a low level of user control, so it's easy to embed the map in a page, widget, or even an info pop-up. It's not necessary to provide the ability to change the appearance of a reference map. Because interactions are usually rare, it's okay to forgo map tools and fancy interaction; ToolTips and hover effects are as far as you should go. Let the content of the page drive the map extent if possible.

Consider adding simple and effective cartographic representation with few distractions. Match the cartography with the branding, style, and color of the rest of the app.

EXAMPLE

Airborne Snow Observatories Inc. (ASO) provides basin-scale inventories of mountain snow water equivalent and snowmelt runoff forecasts with unique remote sensing and physical modeling. Its snow water resource monitoring and management app (figure 2.9) allows users to download and analyze the modeled data to support their critical water management tasks. The app follows the **reference map** layout on the **landing page**. The small map shows available basins. Users can click on a basin that opens an **info pop-up** from which they can select to open a panel to download the data or enter the dashboard of the basin.

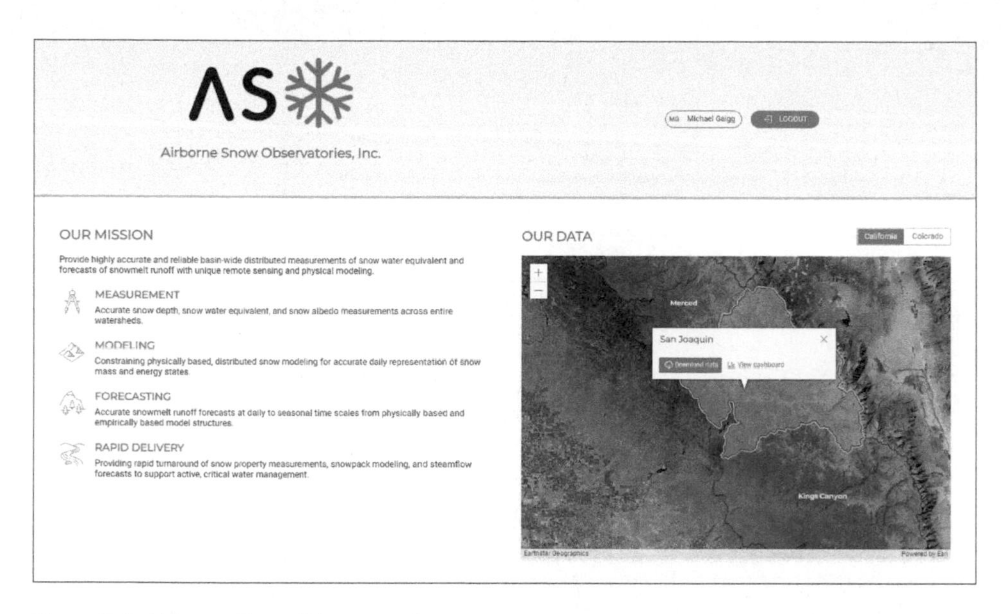

Figure 2.9. Airborne Snow Observatories snow water resource monitoring and management app offers a reference map on its landing page.

Embedded map

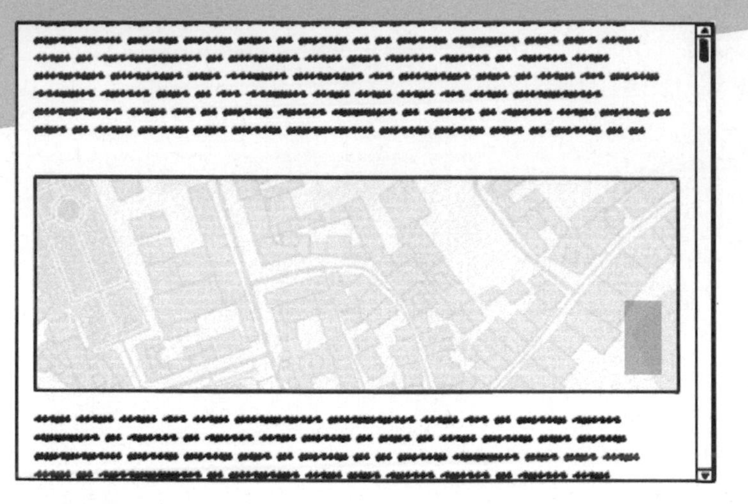

Figure 2.10. Embedded map pattern.

WHAT

In the embedded map pattern (figure 2.10), the app requires an interactive map to perform a simple task, and the map is embedded as part of the page flow. Embedded map layouts are commonly used on multipage sites such as company websites, news sites, blogs, and magazines. They highlight locations, relationships between objects, or statistical data.

WHY

Maps have become an increasingly common way to visualize data on web pages, and it's easier than ever for content providers such as newspapers or blogs to use them. Although site providers often don't have the time, resources, or technology to create their own data and maps, they rely on third-party providers to produce this content and then simply embed it in their pages.

WHEN

In embedded map layouts, the spatial content becomes part of the article or story and supports basic interactions, such as displaying an info pop-up with additional information about a place, the ability to swipe between two data layers, or calculating route directions. A news site may want to show the location of a recent earthquake as part of its reporting, or a food blog may describe native dishes by highlighting their country of origin on a map.

Another popular use that relies heavily on embedded maps is magazine-style layouts that tell immersive digital stories. The location of events, places of interest, or movement along routes or continents are crucial parts of the big picture.

HOW

Different from a **reference map**, an embedded map is typically larger, often spanning the full width or even full height of a screen and responds to scroll interactions and swipe gestures. These interactions may trigger actions that tie content and the map display together. For instance, as users scroll down the page, a map comes into view, gets locked into place, becomes active, and optionally media is scrolled either in a side panel or on top of the map.

Cartography and map style are important factors that draw users to look at the map so they can instantly understand key elements. These elements are especially important for sites with high brand awareness, so try to match the color scheme and typography of the layers to the theme of the site.

Work around the following pitfalls to avoid gesture ambiguity and a debilitated experience:

- Unwanted panning behavior can occur when site visitors use their mouse wheel to scroll through the page, so disable the scroll zoom functionality. Other possible solutions include disabling interactivity with the map until the user clicks it or requiring the user to use two fingers or Ctrl + Scroll to zoom. Both approaches are less common and require further instructions.
- If the map takes up most of the screen, users may not realize that they can scroll the page. Instead, make the map less than the full height of the current viewport and provide cues that scrolling will reveal more content.
- Embedded content has the habit of breaking responsive layouts because of the fixed dimensions of the parent container. Whenever possible, use techniques to resize containers when the dimensions of the browser window change.

Externally hosted maps are often embedded in the app using containers such as inline frames or the HTML canvas element. The provider of the map constructs this HTML code so that the author of the app can embed the control through simple copy-and-paste code snippets. In these cases, the cartography of the embedded map will often not match the style of the page.

Embedding a live map is not always the best approach and depends greatly on the nature of the app. For simple tasks such as highlighting the location of an address, consider using a static map image or a **reference map** instead.

EXAMPLE

An ArcGIS® StoryMaps℠ story is a great example of a site that uses embedded map layout. *The Secret Life of Bridges* (Figure 2.11) looks at the state of America's unsung infrastructure heroes. As the user scrolls through the story, media content such as images and maps enriches the storyline and provides context. The author of the map created a custom **basemap** whose background blends into the site's background, and the color palette of the map matches the accent colors in the paragraphs and tabs.

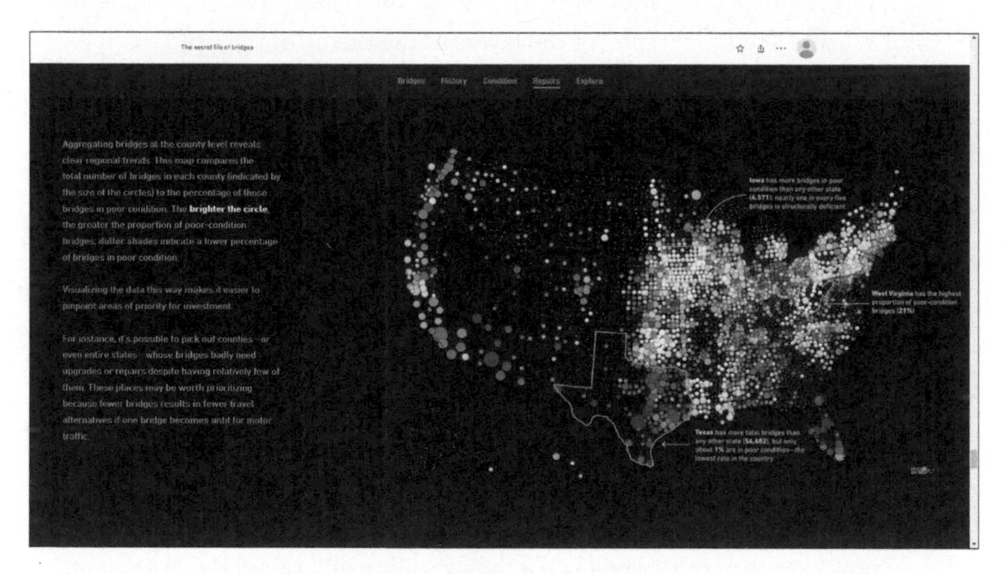

Figure 2.11. The story *The Secret Life of Bridges* uses the embedded map layout to include maps in the app.

No map

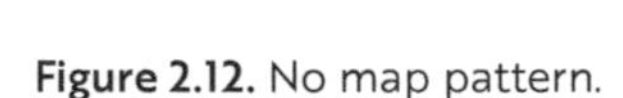

Figure 2.12. No map pattern.

WHAT

The **no map** pattern (figure 2.12) does not use a map. It is used when the task doesn't require a map but still uses the full power of geographic information systems (GIS) in other ways, such as through charts or a dashboard.

WHY

Of all the UI components, dynamic maps fall into the category of "difficult to use." Not everyone can navigate multiscale, data-rich dynamic maps, let alone understand how to read and interpret the spatial content presented. It takes a lot of effort to design dynamic maps well, and thus most maps are far from perfect. Making maps accessible for people with physical or situational disabilities is a constant challenge, as well.

Factoring it all in, it is sometimes a better choice to not show a map but use the available screen real estate otherwise. Often the core value of an app lies in the structured output of the data and can be brought to the forefront in the form of a dashboard instead. This works well when the analysis and the geographic area are well known and don't change. It is then possible to gather and analyze spatial information without the need to display it on, and interact with, a map.

WHEN

Use the no map pattern when displaying a map doesn't add value or is distracting to the task. Location-based intelligence is common, valuable, and can often be represented without a map. Good examples are dashboards that report performance indicators or location services. No map is also recommended when the spatial analysis can be performed and presented more easily without difficult map interactions.

HOW

Design your app without taking up space on the screen for a map. Present the spatial information in an alternative form such as a dashboard with KPIs and charts. Remember that presenting the user with a mapless interface doesn't prevent you from using a map elsewhere in the app. Create separate pages that host a map and provide navigation such as hyperlinks or tabs to access them.

EXAMPLE

The Bin Collection Canberra app (figure 2.13) is a demo app that shows how location analytics can be used without a map. The only input needed from the user is their location, which is provided by typing their address in the **location finder** or clicking the **locate me** button. The app uses that location information to run a query and report back when the next rubbish, recycling, and green waste pickup days are scheduled. These simple answers are what most users need—mission accomplished. Having a map would unnecessarily complicate the interface and provides little additional value to the app. Quite the opposite, map authors would have to solve new questions that arise: How should we visualize multiple trash truck routes across the city in a meaningful way? Should we aggregate the routes into polygons such as service areas? How do we deal with overlapping service areas? Ask yourself, what do your users really need, and resist the urge to add a map at all costs and just because you can.

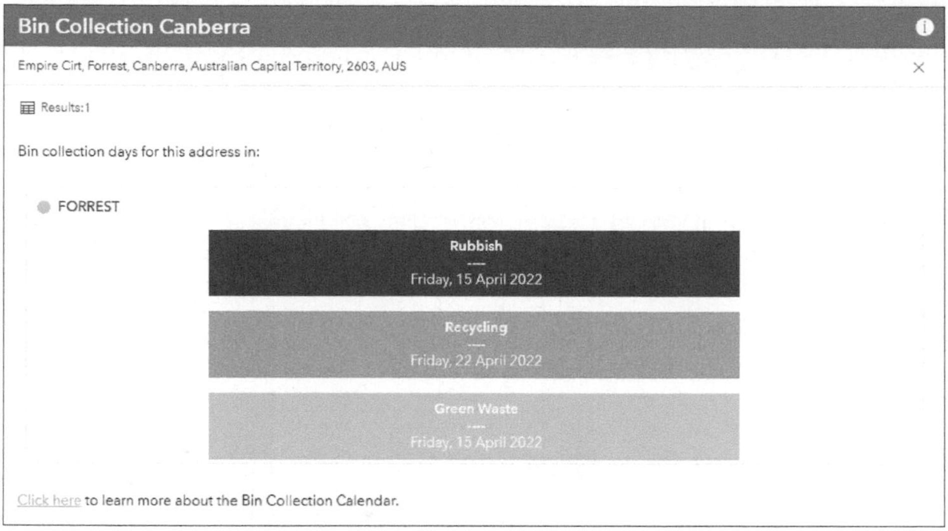

Bin Collection Canberra

Empire Cirt, Forrest, Canberra, Australian Capital Territory, 2603, AUS

Results: 1

Bin collection days for this address in:

FORREST

Rubbish

Friday, 15 April 2022

Recycling

Friday, 22 April 2022

Green Waste

Friday, 15 April 2022

Click here to learn more about the Bin Collection Calendar.

Figure 2.13. Bin Collection Canberra shows bin collection days for a particular address.

Focus on the subject

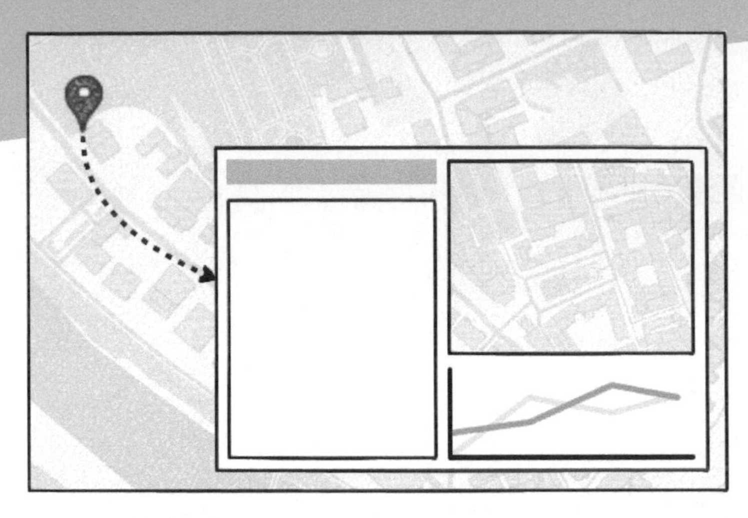

Figure 2.14. Focus on the subject pattern.

WHAT

Focus on the subject (figure 2.14) is a pattern with many elements that fosters an interaction method allowing the user to switch their focus or center of interest or activity from a generic display to a specific, detailed, and focused screen.

WHY

Selecting the correct layout is a challenging design task. Each layout pattern has a purpose and benefits but also comes with shortcomings. Builders must decide early on the right layout and how to arrange the necessary elements on the canvas. Plenty of thought goes into the UI design, and just when every piece seems to be in place, the user clicks a map marker and the lens of interest changes. Instead of exploring a map full of objects, the focus shifts to a particular object that becomes the subject of interest. This new view requires a different kind of layout, so the app must toggle between the two layouts, often **full map** and **reference map**. The focus on the subject pattern doesn't constrain the app from using another layout when necessary, regardless of the initial layout used.

WHEN

Use the focus on the subject pattern when the current view or layout doesn't provide enough space to display all the information for the subject of interest. This is often the case for full map layouts that need to show details for a data point and in which an info pop-up doesn't suffice. Toggling the view to a reference map layout on a separate page or inside a modal window is a great way to show details about the subject of interest.

HOW

Provide a switch so that users can toggle to an alternative layout. This switch can be a hyperlink that navigates to another page or a button in an info pop-up that opens a modal window. The new layout then has enough space to show additional properties, tables, charts, or even a small reference map to illustrate the location.

Ensure that the change in layout is seamless and logical, and provide smooth transition during navigation with a clear way to go back. Make it easy for users to dismiss the selection, for instance, with a Close button in a modal window.

Users don't like change or learning an app with new functionality and abilities. Provide familiar UI elements to avoid annoyances and disruptions. Limit the amount of layout changes by using the focus on the subject pattern only when needed. Consider fitting the content into already available containers, such as a side panel, or use the info pop-up or info panel whenever possible.

EXAMPLE

The Geospatial Insurance Consortium (GIC) is an insurance industry consortium that provides its members with advanced imagery and data to enhance the underwriting assessment process. The GIC Virtual Inspection Studio app uses focus on the subject to allow virtual adjusters to expedite claims and improve fraud detection following disasters. Users enter the app through a UI that is discussed in more detail in the location list pattern (chapter 6). This initial page shows the insurance claims in a list and on the map. When the user selects a claim from the list, the app leads them to the details page (as shown in figure 2.15) with a split-screen before-and-after imagery viewer that shows a property before damage from a disaster and the conditions immediately after. This navigation behavior is different from using an info pop-up or the list and details pattern. The reason for showing a new page that focuses on the subject instead is that the visual inspection requires space to compare high-resolution imagery so that adjusters can focus on the task to complete the claim assessment for an individual property.

Figure 2.15. The Virtual Inspection Studio app uses the focus on the subject pattern to allow virtual insurance adjusters to visually inspect a property.

References

"Mapping APIs and Location Services," https://developers.arcgis.com/documentation/mapping-apis-and-services.

Chapter 3
Interacting with the map

Dynamic maps differ from static maps in that they allow users to manipulate the extent and content of the map by working with it. Fittingly, these maps are therefore often referred to as interactive maps and serve as the main building block for rich web apps. This chapter focuses on the two main interaction techniques with maps: navigation and information retrieval.

Navigation

Interactive maps are difficult in many respects, and people struggle to use them more often than you'd think. Building your next imaginative and elaborate map app is therefore rewarding and challenging at the same time. The reward is having a rich UI component that allows exploration and manipulation of multiscale and multifaceted spatial data. The challenge is navigating and understanding this complex and vast body of data. What makes matters worse is that so far there hasn't emerged a comprehensive and agreed-on set of rules or guidelines on how to interact with a map across devices and services. Table 3.1 shows the four main map interactions and their corresponding commands. Depending on the choice of your mapping platform, the interaction events might differ, but the good news is that, in most cases, you don't need to adjust.

Additionally, most maps also show the explicit **zoom control** component for zooming in and out, and some mapping platforms allow rectangle selection while pressing and holding one of the control keys—Shift, Ctrl, or Alt—to zoom into the desired area.

Table 3.1. Map controls for interacting with a map

Command	Mouse	Keyboard	Gesture	Game pad
Zoom in	Double-click Scroll forward	Arrow keys	Two fingers pinch in	Left trigger
Zoom out	Ctrl + Double-click Scroll backward	+	Two fingers pinch out	Right trigger
Pan	Drag	–	Drag with one or multiple fingers	Left stick
Rotate	Right-click + drag	A (counter-clockwise) D (clockwise)	Two fingers rotate	Right stick

Information retrieval

Whereas static maps are thought of as what you see is what you get (WYSIWYG), interactive maps are only the gateway to accessing large collections of information. Users have become so widely accustomed to clicking in a map to retrieve more on-demand information that omitting this functionality might appear to many as if the map were broken. The process of getting to additional information that is currently not visible is called information retrieval and can happen in two ways:

- **Navigation:** Zooming between scales, rendering additional layers at different granularity
- **Trigger events:** Revealing more details about features on the map

As you will learn in this chapter, **markers** are the main pattern to retrieve information, which can be displayed in either a **MapTip**, an **info pop-up**, or an **info panel**. Each of these patterns is described in more detail later. Their common properties are trigger method, position, dimensions, content, and actions. Table 3.2 compares those properties.

Mobile devices don't have hover events that trigger MapTips and have a different input mechanism called gesture events. Table 3.3 describes the information retrieval patterns on mobile devices in which hover state doesn't exist and screen space is limited.

Table 3.2. Usage of information retrieval patterns

Pattern	Trigger			Position	Dimensions	Content	Actions
	v1	v2	v3				
MapTip	Hover	Hover	n/a	Top/ Bottom/ Left/Right	Small	One or two attributes	n/a
Pop-up	Click	n/a	Hover	Top/ Bottom	Medium	Title, description, few attributes, actions	Start task, share, like, rate, zoom to, etc.
Panel	n/a	Click	Click	Left/Right Docked	Full height	Long, rich, interactive	Further interactions on map okay

Table 3.3. Usage of information retrieval patterns on mobile devices

Pattern	Trigger	Position	Dimensions	Content	Interaction
Pop-up	Tap	Top/Bottom Docked	Small	As little as possible	Few buttons
Panel	Tap	Cover	Full height & width	Scrollable	More buttons okay

Patterns

Adding interactivity to the map enables users to take control of the data display and empowers them to take charge of the content. The patterns in this chapter belong to the two common groups of interaction techniques found in interactive maps: navigation and information retrieval.

Navigation patterns:

1. **Zoom control**
2. **Marker**
3. **Rich marker**
4. **Visited marker**

Information retrieval patterns:

5. **Info pop-up**
6. **Info panel**

7. MapTip
8. Callout (www.mapuipatterns.com/call-out)

Patterns that directly influence how users interact with the map are related to basemap layers, as follows:

9. Basemap (www.mapuipatterns.com/basemap)
10. Basemap toggle (www.mapuipatterns.com/basemap-toggle)
11. Basemap gallery (www.mapuipatterns.com/basemap-gallery)

Patterns to compare content are as follows:

12. Swipe (www.mapuipatterns.com/swipe)
13. Side by side (www.mapuipatterns.com/side-by-side)
14. Small multiples (www.mapuipatterns.com/small-multiples)

Patterns for common input and output tasks are as follows:

15. Data extract (www.mapuipatterns.com/data-extract)

The following visualization patterns describe different views of a map:

16. 2D (www.mapuipatterns.com/2-dimensional-2d)
17. 3D

Zoom control

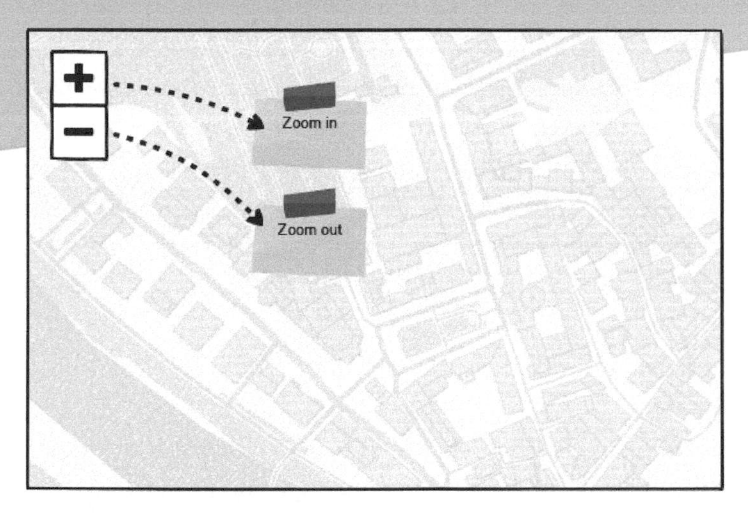

Figure 3.1. Zoom control pattern.

WHAT

The zoom control (figure 3.1), sometimes called zoom buttons or map navigation, is the most important control of an interactive map. Zoom control is used when the current viewport doesn't show the appropriate level of detail, and the user needs to see a larger or smaller map scale.

WHY

Although modern map apps provide alternative means of zooming in and out through gestures and mouse events, many people, especially on touch devices, still prefer the more precise navigation technique of clicking buttons over pinch-zoom or the use of a mouse wheel.

The main actions on the zoom control, zoom in and zoom out, are represented by the plus and minus signs, respectively. The origins of this association stem from the increase and decrease in scale factor: zooming in will make the map scale larger, and zooming out, the map scale smaller. Even though most people don't know about the origins of this usage, and some may argue that plus and minus don't visually resemble their functions enough, users have grown accustomed to this representation. Apps should leverage this familiarity and thus follow this convention.

WHEN

Any interactive map requires a means to increase or decrease the scale, sometimes referred to as zoom level. Zooming in and out is the foundation for navigating between levels of detail shown in the map. It allows users to focus and change their center of interest. Apps may start with an aggregated view of the data, such as insurance claims, available grants, or broadband availability at small scales. As users zoom in to medium scales, the aggregation turns off and starts to show the available data points. At very large scales, the extent shows only one or a few features.

HOW

Most mapping platforms provide the zoom control out of the box. It consists of two vertically stacked buttons to zoom in and zoom out depicted by the plus and minus signs, respectively. Sometimes the buttons surround a slider component to speed up navigation between scales. Adding ToolTips to describe the button's function is a best practice.

The zoom control is usually located in the upper-left corner of the map, but any other corner works as well. The decision about corner placement should be driven by its hierarchy of importance and support a well-balanced layout. For instance, if the upper-left corner is occupied with a **search** that is the focus of the app, placing the zoom control in the upper-right or lower-right corner provides a good counterweight. Removing the zoom control altogether isn't recommended although not wrong, especially for apps with a narrow and well-known user base or on mobile devices with limited space.

The transition between zoom levels should be smooth with a cinematographic motion indicating the change. Sometimes the new image is a blurry version of the previous image until the new imagery has loaded, although this is less of an issue when using vector data. Zooming should center the map at the same midpoint.

In case your app has more than one map, each dynamic map that isn't synchronized with the extent of another map requires its own zoom control.

EXAMPLE

The ArcGIS Vector Tile Style Editor (VTSE, figure 3.2) is an online tool for styling vector tile layers such as basemap layers. The editor interface shows four interactive maps to preview styling changes at different zoom levels simultaneously. Because all four maps are independent of each other, they each need their own zoom control to indicate that they are indeed interactive maps and not just preview images. Somewhat unique to this app is that the map scale is important for styling the layers that make up a basemap layer at each of the 23 scales. For that reason, each map shows a number beneath the zoom control that indicates the current zoom level.

Vector Tile Style Editor lets you stylize vector tile layers at different scales. Learn more at developers.arcgis.com/vector-tile-style-editor.

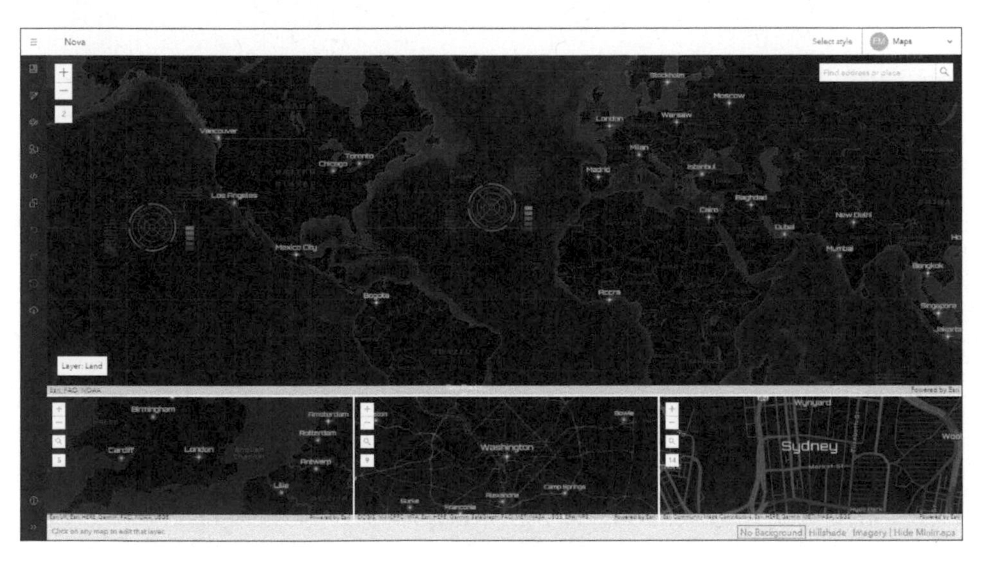

Figure 3.2. Vector Tile Style Editor shows zoom control for each interactive map on the page.

Marker

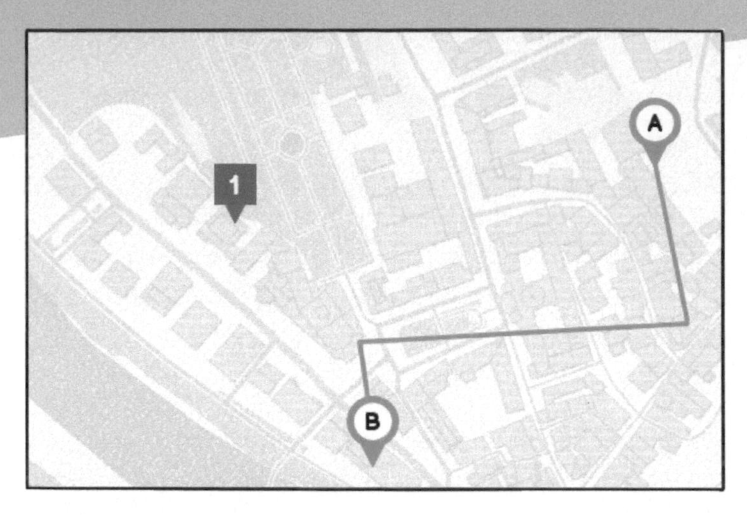

Figure 3.3. Marker pattern.

WHAT

A marker (figure 3.3), sometimes also called a map pin, identifies a location in the map. This location can be a place, point of interest, or any other object.

WHY

Adding markers is one of the most important and powerful functions of a map because it helps pinpoint the location and establish a connection to reveal additional content. Markers are strong focus points with a high capability of being interacted with. They typically trigger MapTips, info pop-ups, or info panels.

WHEN

Markers are used to highlight a special location on the map, usually after input by the user such as clicking a search result, using the location finder, or selecting a point feature. In this sense, markers are different from point symbols because they indicate points that were selected by the user or stand out through their importance to the subject, as shown in a store locator, for instance.

Route positioning uses markers to indicate the start and end points of the route. The location list uses markers to establish a unique relationship between the item in the list and the point on the map. Locator apps such as hotel booking sites, store locators, or site finders may choose to display points using markers instead of simple point symbols.

HOW

Markers differ in color, shape, and size but are typically graphics. As a convention, the upside-down, teardrop-shaped marker indicates an exact location selected by the user. This user input could be clicking a feature, performing a search, using the location finder, or selecting a way point of a route. Alternative marker symbols can take any form or shape, such as pushpins, circles, or rectangles, and may include numbers or letters to reference content displayed in a list.

Markers must be interactive and support click or tap and optionally hover or drag interactions in the following ways:

- Click or tap should open an info pop-up with additional information about the location, such as title, description, photos, reviews, or action buttons.
- Hover opens a MapTip displaying short and informative information about the marker.
- Drag is used to adjust the location. On mobile devices, the marker may be placed in a fixed center position of the map, allowing users to drag the map underneath instead of dragging the marker itself.

The decision whether to use markers or regular point symbols should be based on the following:

- **Density:** Marker. Data with few overlapping points lends itself better to using markers.
- **Information retrieval:** Marker. Necessity to reveal more detailed information may indicate the need for markers.

- **Subject matter:** Points. Indicating physical locations works better with points.
- **Classification:** Points. Points that don't need to be symbolized based on an attribute value are preferred but might still work if classification is Boolean, such as on/off, available/not available, or found/not found.

After a marker is placed, the map zooms to and centers on it; adding a tiny bounce animation is nice, too. The exact zoom level depends on the purpose of the app. A parcel viewer app may zoom to a large scale showing the property at that address almost entirely, whereas a store locator may zoom to a smaller scale that includes multiple stores close to that address.

EXAMPLE

The Commercial Real Estate Site Selection app (figure 3.4) allows users to compare multiple retail, office, industrial, or mixed-use sites side by side. The app uses ArcGIS GeoEnrichment℠ Service to enhance existing data with additional location-based information such as demographics. When a user accesses the app, they see a list of sites and an interactive **reference map** with graduated symbols. Larger symbols indicate that more sites exist in this area. The items in the list have an icon button to add the site as a comparison card to the main content area. Each comparison card shows basic information, including the location and geoenriched data about the site, such as population, housing, income, and education. A unique feature of this app is how it uses a static map with a **marker** on top to communicate the site's location. The boundary around the marker shows the area an average person can walk in 10 minutes, which is used as input for reporting the site statistics.

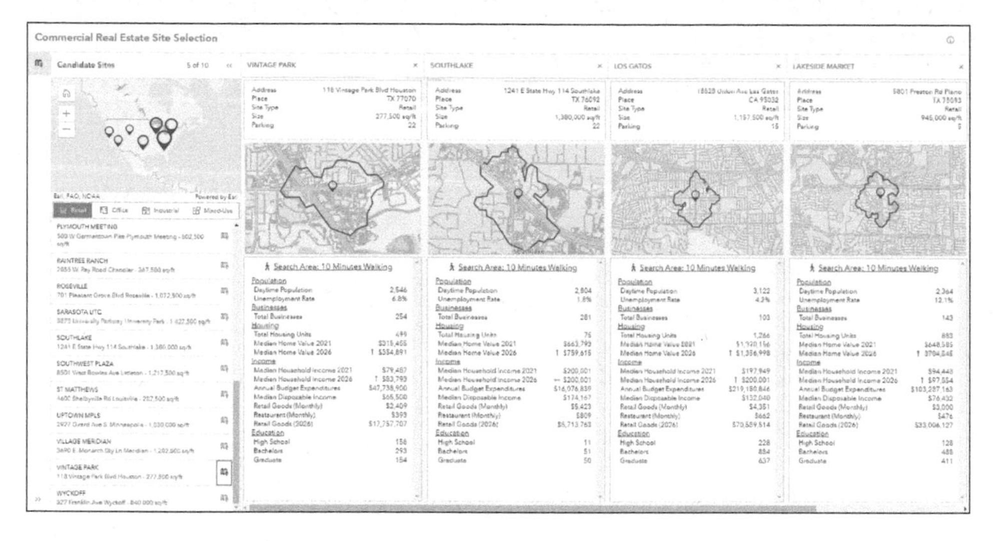

Figure 3.4. Commercial Real Estate Site Selection uses markers to show the exact location of selected sites.

Rich marker

Figure 3.5. Rich marker pattern.

WHAT

The **rich marker** pattern (figure 3.5) extends simple markers on the map to show a key metric, performance indicator, or other important information.

WHY

Sometimes the color, size, or shape of a **marker** is not enough to depict the information a user is looking for. Although this information can be accessed through the **info pop-up** that appears after clicking a marker, a user would greatly benefit by not flipping between individual markers and having to recall their content.

WHEN

Use a rich marker when the content of a set of markers is important for decision-making and the user benefits from seeing this information immediately. A **store locator**, for instance, may want to enrich individual stores by showing the distance or duration from a user's location, or a hotel booking site may want to indicate the price for the cheapest room in addition to the place itself. A sophisticated rich marker might even display actions that users can directly interact with, such as share or like, without the need to open an info pop-up first.

HOW

Before adding rich markers to your map, consider that they tend to clutter the UI with additional content and can easily lead to information overload and overwhelm your user. As a result, rich markers work best when the result set is limited to a small amount and without overlapping points. In the case of overlapping markers, consider implementing a mechanism that brings partially hidden markers to the foreground on a hover event. Adding subtle hover effects to highlight the border or background or gently increasing the marker size are helpful feedback mechanisms.

Make the rich markers larger than simple markers so that additional content can be displayed inside. Show metrics or indicators that are directly related to the context of the task. Numbers must be meaningful and equal in scale and unit; indicate the unit, such as miles or degrees, if not immediately evident from the UI. Users still expect to see an info pop-up upon clicking, so it's a good practice to show the pop-up even though the same information has already been pulled to the forefront through the rich marker.

EXAMPLE

This mock-up of a vacation rental (figure 3.6) imagines an online marketplace for travel lodging. Its users are vacationers who want to find vacation rentals based on location and travel period. Price is the key differentiator, and that's why the map markers show the average nightly price inside a rich marker instead of using static markers or color-coded points with a low/medium/high classification.

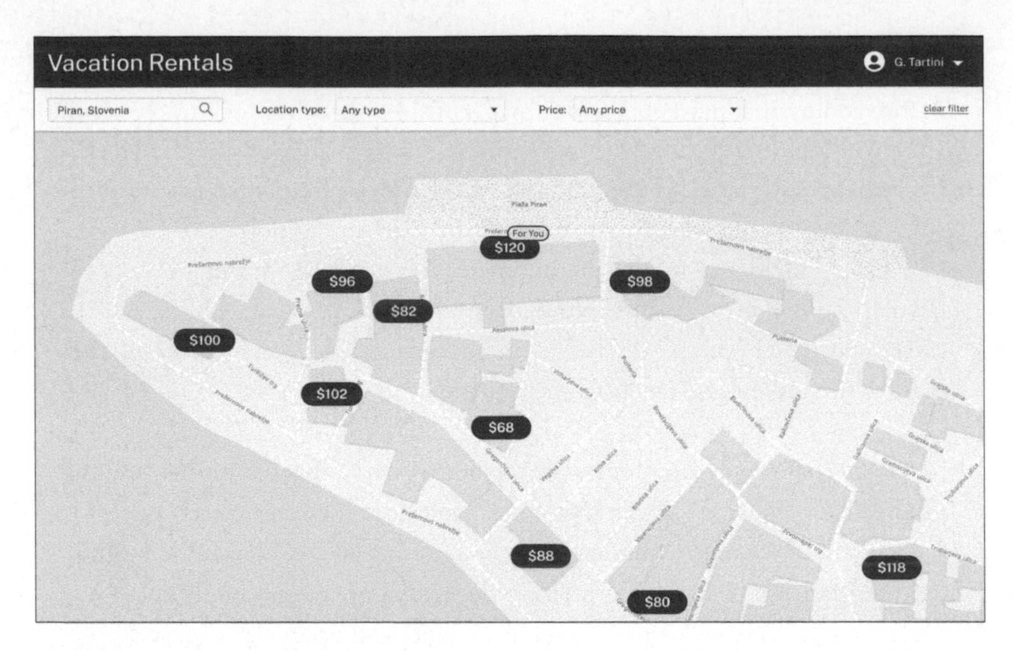

Figure 3.6. Vacation Rentals mock-up enriches its map markers with prices to avoid extra data retrieval.

Visited marker

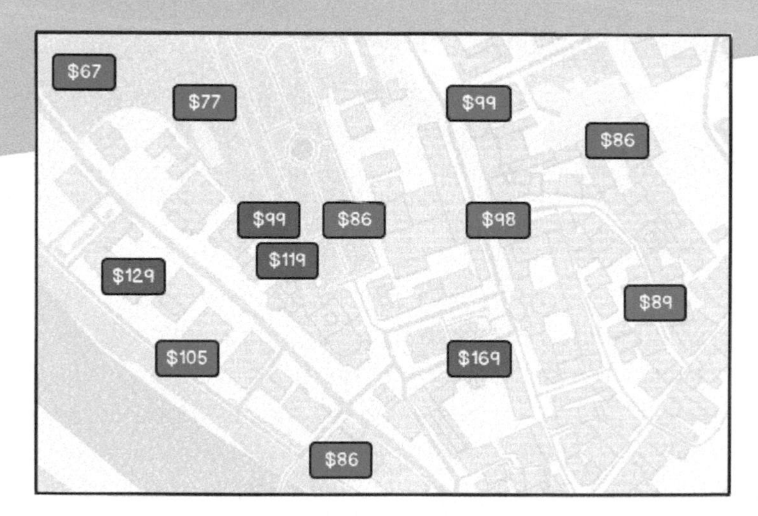

Figure 3.7. Visited marker pattern.

WHAT

The visited marker pattern (figure 3.7) uses markers symbolized to show users which marker has already been visited, helping them avoid revisiting the same info pop-up repeatedly.

WHY

A pattern that has been around since the early days of HTML is to change the color of hyperlinks that have been visited previously. This is considered a usability best practice and helps users recall which hyperlinks have been visited in the past so they can either be revisited or excluded the next time. The same principle can be applied to map markers, which frees users from unintentionally revisiting the same pop-up repeatedly. This pattern is especially important when overlapping markers change their display order and visited points come to the forefront again. Users will end up seeing the same pop-up again and again and as a result give up faster out of frustration.

WHEN

Using the visited marker pattern helps users make faster decisions by increasing their focus on specific properties. Use it in single-purpose apps that are targeted to specific topics such as hotels or restaurants in which users explore an area of interest to make a booking or reservation. What appears to be a small detail might set your app apart from similar apps in the marketplace.

HOW

Show visited markers with a different background color, typically a shade of gray that is distinct from the marker's primary color. After the user has clicked a marker to trigger the **info pop-up**, the previously colorful marker turns gray. This subtle change allows users to quickly see which markers have already been explored and reduces their anxiety and cognitive load to remember and recall. Changing the font color or weight is an acceptable alternative but less effective.

To comply with accessibility guidelines, remember to choose color/label combinations that provide a contrast ratio of at least 4.5:1. Use a contrast ratio tool such as https://contrast-ratio.com to find accessible foreground/background combinations. Alternatively, consider using a different shape or outline to avoid relying on color alone.

EXAMPLE

Recognition over recall is Vacation Rentals' approach to using the visited marker pattern (figure 3.8) to indicate vacation rental markers that have already been visited. This feature reduces the cognitive load of having to remember already seen properties, empowers users to feel as if they are in control, and speeds up the booking process, which supports the app's business model.

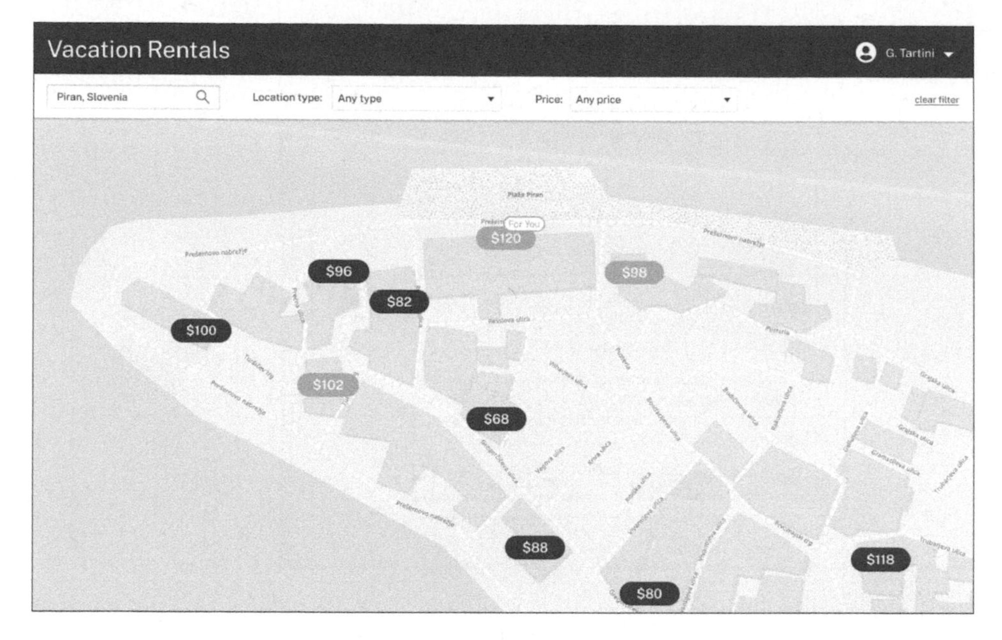

Figure 3.8. Vacation Rentals uses a lighter color to indicate visited markers.

Info pop-up

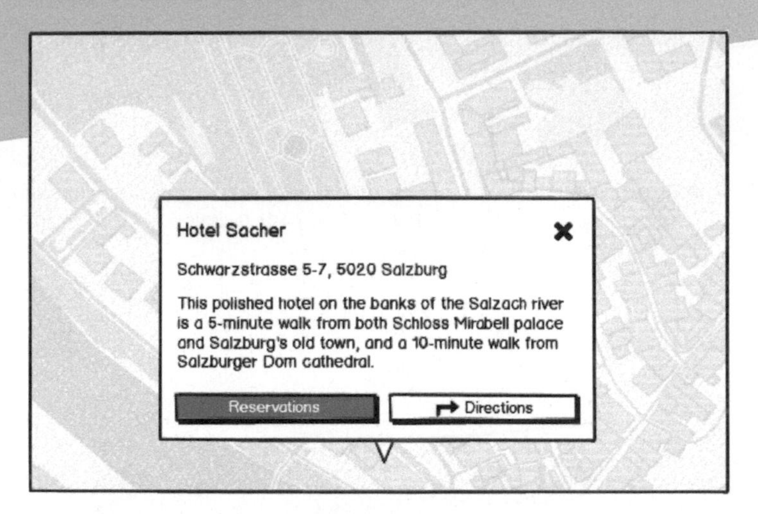

Figure 3.9. Info pop-up pattern.

WHAT

The info pop-up pattern (figure 3.9) employs the primary vehicle for accessing content and retrieving information about features on the map. The info pop-up, sometimes also called info window or just pop-up, is the most recognizable object of a map app but doesn't receive the attention it deserves.

WHY

Builders tend to be content with the default display of attribute value pairs and overlook the potential of knowing the user's intent while they interact with a feature. As such, data retrieval by itself shouldn't be the end of a task. It can be the beginning of a workflow that uses the selected location or feature as input for getting route directions or making a reservation.

WHEN

Users have become so accustomed to opening a pop-up by clicking in the map that any layout, except the reference map, should provide pop-ups. Furthermore, pop-ups are the primary method for accessing additional data about objects on the map, so it makes sense to provide space inside them for information and actions. When users trigger a pop-up, their intention was most likely to interact with this feature, so any actions provided inside the pop-up, such as sharing, rating, deleting, or editing, are more likely to be used.

HOW

When someone clicks in your map, they should be rewarded with a bundle of useful information about that location. To make this happen, you need to consider four aspects: how the pop-up is triggered and closed, its position and dimensions, the content, and the actions. Most mapping frameworks take care of trigger interactions and positioning, but it's nevertheless important to discuss them here.

The simplest way to trigger a pop-up is to click a marker or feature. Alternatively, opening can be triggered after selecting a search result (see **search** and **location finder** in chapter 1, "Getting Started with Design") or clicking a list item (see **location list** in the "Marker" section earlier in this chapter). Indicate the selected location by changing the feature symbol to the selected state, placing a marker on top, or replacing it with a marker altogether. Users can close the pop-up by clicking the *X* symbol or a Close link in the upper-right corner. In recent years, more apps omit the explicit option to close the pop-up in favor of clicking elsewhere in the map, which will open another pop-up or close it in case no information exists. It's important to note that only one pop-up can exist at a time, which means that when a new pop-up opens, the currently active pop-up must close.

Usability guidelines have changed over time but current suggestions for positioning are to not pan the map to place the pop-up above the marker but find another position below or to the side. This helps users keep oriented without losing their sense of the marker location. On mobile devices where space is limited, dock the pop-up to the bottom or even the top of the screen. Don't make the pop-up floating or draggable. In cases where map content must not be hidden, the **info panel** is the better choice. Always try to minimize the space that a pop-up requires by optimizing its content and avoid the need for scrolling whenever possible.

You will find that the main work for creating useful pop-ups is deciding on the information that is shown in the content area. **Info pop-ups** can contain photos, videos, ratings, charts, or even **route directions**, but aim to avoid information overload. A common technique to hide additional content until the user explicitly asks for it is progressive disclosure. As a result, the interface will appear less cluttered and will allow users to focus on exploration and navigation until the desired location is found. Avoid showing meaningless table columns such as ObjectID, Federal Information Processing Standards (FIPS), or other codes and abbreviations, but instead compile and compose the content into human-readable language and structure. Use visual hierarchies to structure the content. Although you are using a map, there is rarely a good reason to include latitude and longitude fields.

Provide meaningful actions in the form of buttons or hyperlinks that allow users to use the current object as the starting point for further analysis or exploration or to share, like, rate, and zoom to.

Especially beware of the antipattern of showing an empty pop-up with a message that reads, "No information found." Either decide on not showing anything or use the input coordinates as an opportunity to start a search instead. In the latter case, reverse geocode the geographic coordinates to a human-readable address, or place and show a pop-up that includes one or more of the following actions: add missing place, route from here, route to here, report a data problem, measure distance, or search nearby.

EXAMPLE

Space Planner for ArcGIS Indoors™ (figure 3.10) is a browser-based app that allows users to plan occupant activity in indoor spaces, including assigning occupants to individual spaces or activity-based work areas called hotels or hot desks. Users start by creating a plan to evaluate several assignment scenarios. The scenarios are adjusted throughout the planning and review processes before the plan is chosen that best fits the organizational needs. The new assignments can then be exported for use in other Indoors apps such as booking meeting rooms or office hotels. The main interaction mechanism to view data about an office, office hotel, hot desk, or person is the info pop-up. This pop-up shows the title, important metrics, current and planned assignments, properties, and related items. It also allows users to assign and remove entities through actions available inside the pop-up. Although the same actions can be accessed elsewhere in the app, having them readily available feels intuitive and saves time.

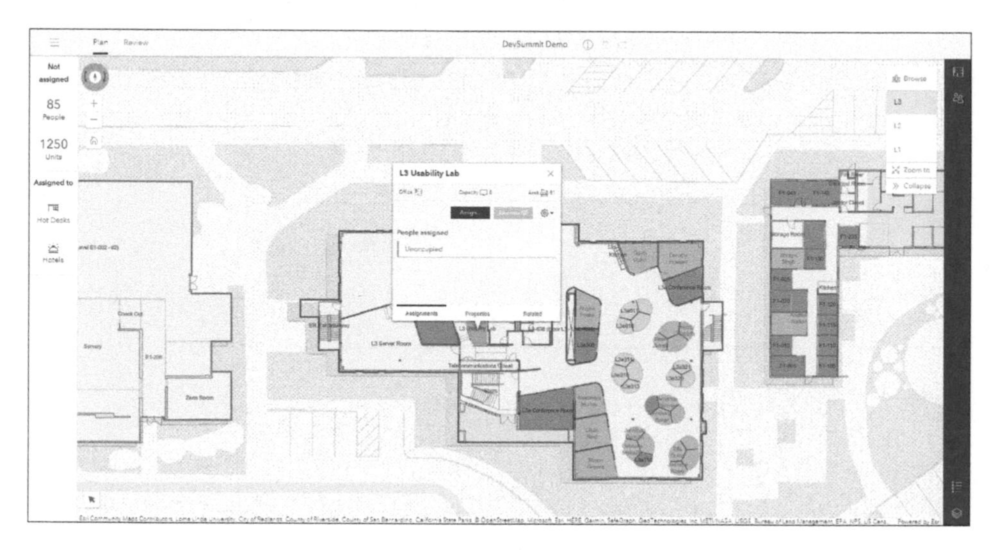

Figure 3.10. Space Planner for ArcGIS Indoors makes extensive use of pop-ups to reveal information and let users act on the content.

Info panel

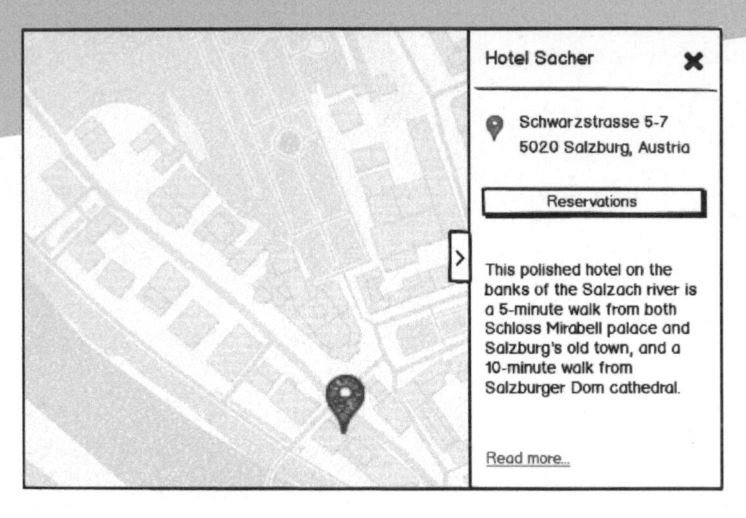

Figure 3.11. Info panel pattern.

WHAT

The info panel pattern (figure 3.11) provides for the display of long or interactive content that doesn't fit into the small space of an info pop-up or that must not obstruct the map.

WHY

The info panel is, like its cousin the info pop-up, an important UI component to display information about an object on the map. The reason that builders may choose a panel over a pop-up is that it provides a reserved area of the layout to display long, rich, or interactive content, whereas the pop-up has space only for a title, a few attributes, a short description, and some action buttons. The info panel is especially useful when further interactions with the map are required and potentially overlapping pop-ups are undesired or inefficient.

WHEN

Use the info panel when the content to be displayed is long, interactive, requires further interaction with the map, or must not cover parts of the map. Apps in the safety and security space often have a hard requirement to avoid obstructed content because it might hide potentially important data, such as incidents, emergency vehicles, or 911 calls. Enterprise apps geared at expert users often tie advanced mapping tools and workflows directly into info panel content, and apps that use the list and details pattern (see chapter 6, "Building Single-Purpose Apps") may use the available panel to show details of the selected object.

HOW

Provide a panel on the left or right of the map so that it serves as the container for the content. On mobile devices, the panel should take up the full screen and operate like a modal window. It's advisable to make panels closable or collapsible to avoid taking up map real estate when they aren't needed. After a feature is clicked, the app adds a marker at that location and slides open the panel without centering the map on that marker. This animation will cause the map extent to change by clipping equal parts of the map on the left and the right while preserving the map center. This may lead to the less common case that the selected marker is on the portion of the map that got removed, but users understand that a simple pan will move the marker in sight again and still prefer this behavior over automatic panning, zooming, or clipping.

Remember that the info panel effectively replaces the info pop-up, so consider adding MapTips to provide immediate feedback. Some apps may even display sticky MapTips that resemble an info pop-up and stay in place even after the pointer moves away.

Last, it's important to consider the empty state of the info panel. Design special content that can be displayed while the panel is open but before the user clicks a feature. In this case, the panel can either serve as a place for instructions on how to select something, show a graphic that adds interest to your app, or use the space for other useful content, such as a list of curated locations. If no such use can be identified, keep the info panel closed by default.

EXAMPLE

ArcGIS Indoors software is a complete indoor mapping system for smart building management. Its indoor viewer app (figure 3.12) allows you to find office locations and resources within a building or a site. The app starts by displaying the whole site. Selecting a building lets you peek inside to see its floor plans. As you do that, an info panel opens on the right showing detailed information about the building or feature. Within that panel, users can then invoke actions such as routing to other features, booking meeting rooms or office hotels, calling or emailing occupants, or logging issues and work orders. The info panel is slightly offset from its surrounding borders to distinguish it from the content panel on the left and indicate that the information shown is related to a selected feature in the map.

Figure 3.12. Indoors viewer adds a marker to the selected feature and opens an info panel on the right side.

MapTip

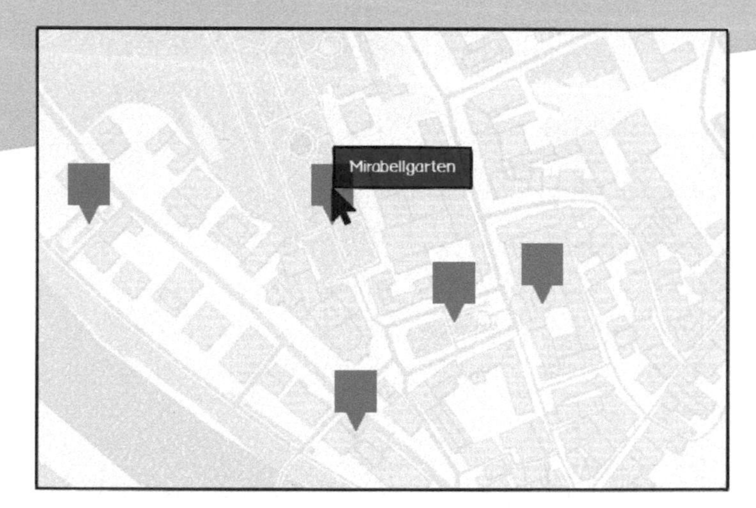

Figure 3.13. MapTip pattern.

WHAT

A MapTip (figure 3.13) is an informative message that appears when a user hovers the pointer over a feature, surface, or raster image. The purpose of a MapTip is to provide helpful content such as additional attributes, data, or images and therefore give users direct access to attribute information for that feature.

WHY

The map provides available content through markers, rich markers, and labels. Additional information can be retrieved through interactions that open sometimes invasive elements such as info pop-ups or info panels. As such, MapTips are useful for showing information that cannot be displayed on a marker or label but that doesn't justify disrupting the user experience by opening a pop-up or panel.

WHEN

Because MapTips are initiated by a hover gesture, they are not accessible on touch screen devices and therefore should be used only to display information that is helpful but not essential to complete a task. An example of a useful MapTip is a tourist map that shows travelers the name of an attraction, hotel, or restaurant. On **choropleth** maps, the MapTip could show information about the hovered geographic area such as total value or even a small pie chart that graphs values by a certain attribute.

HOW

Provide MapTips that appear when hovering the pointer over a feature or tabbing through features using the keyboard. The MapTip should be self-sufficient with brief, helpful, clear, and consistent content. Avoid redundant or lengthy text; for numbers, add units.

Place the MapTip close to the hovered feature and at the pointer's tail or head if possible. To avoid potential confusion, consider adding pointer arrows to indicate the relation of the MapTip to the feature. This is especially important when several nearby or overlapping features may have triggered the MapTip.

The MapTip can show immediately or after a brief delay, which depends on the anticipated usage of the information and weighing the value and amount of information with the annoyance of constant interruptions. For performance reasons, it is sometimes recommended to disable MapTips when they are not necessary, at smaller scales, or when they turn out to be more distracting to the user than helpful.

Users cannot interact with the content of a MapTip because moving the pointer will make them disappear. Therefore, do not provide links or buttons inside a MapTip because users will not be able to reach them. The exception to this rule is when MapTips are implemented so that they stay in place, or are sticky, without immediately disappearing. In this case, they can be treated as if they were **info pop-ups**.

Another aspect of MapTips is that they are triggered by the user, which distinguishes them from similar interactions such as through **onboarding** tutorials, which show tips automatically.

EXAMPLE

Find Your Candidates (figure 3.14) is a web-based app that was used to visualize the US House and Senate races in the 2020 election. It allowed users to find officials running for public office and to learn more about them. The points represented races for the House of Representatives and the photos represented races for the Senate, also highlighting the incumbent congressperson and their challengers. The app used MapTips to show the candidate's name and party affiliation as a quick way to get basic information before the user clicked the candidate to see more details in an **info panel**. The app also featured an **attribute filter** to filter congresspersons by election level, party affiliation, or race and gender.

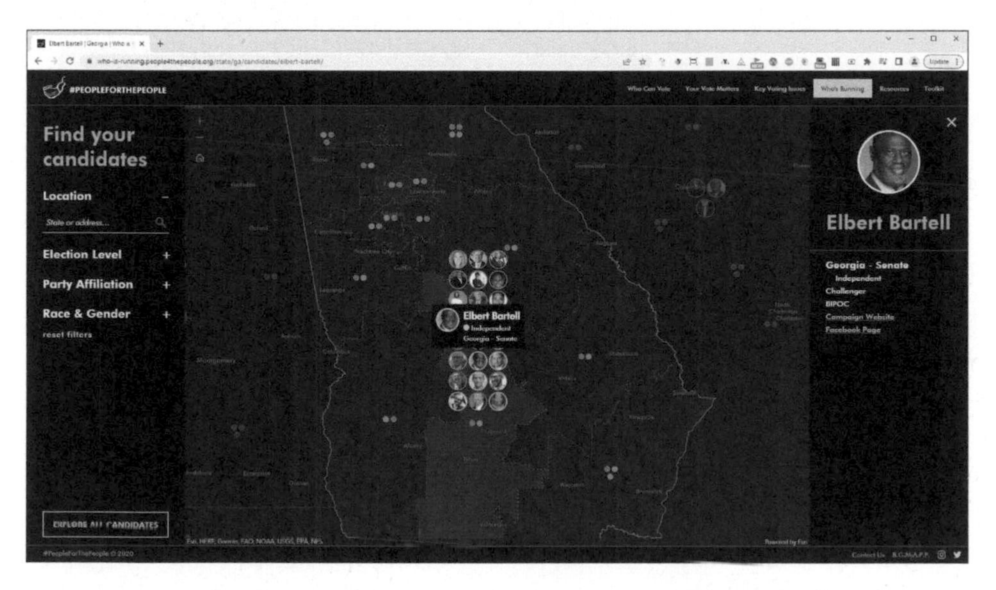

Figure 3.14. The Find Your Candidates app helps users explore information on election officials.

3D

Figure 3.15. 3D pattern.

WHAT

The **3D** pattern (figure 3.15) features maps that provide appealing, realistic views that are great for complex landscapes and vertical information. They provide a real-world view that can be easier to interpret.

WHY

Throughout history, 2D maps have been the main way to depict geography, from drawings on animal skin, papyrus, and paper to digital maps on computer screens. The disadvantage of any 2D map is that it requires a high level of geographic literacy to be interpreted well. 3D maps, on the other hand, have been around for less time and are mostly limited to hand-drawn, artistic renderings of cities or landscapes from a bird's-eye view. Their main advantage is that people can intuitively understand them because they are realistic representations of the world as they know it. For instance, it's easy to identify landmarks on a 3D map or see that an area is a lake and not a forest.

With the advent of 3D information models and faster computing power to render them, 3D maps are now a viable alternative to 2D maps. It's easy to control camera tilt, angle of view, and even lighting. 3D maps have a higher positive appeal because of their realistic representation of area surroundings. This helps users orient themselves and find objects of interest more quickly.

WHEN

Use 3D maps to illustrate a more naturalistic picture of complex landscapes or when the third dimension is an important factor, such as in construction, real estate, or landscape planning. 3D maps are especially useful in situations in which users need to quickly orient themselves, such as tourist maps or virtual tours. Use 3D to provide a common operational picture—for instance, when business operations need to know where things are and how they are connected or analysts and planners need to understand the complete scenario. Overall, 3D maps add realism and are helpful in instilling confidence that things are what they appear to be.

HOW

Start the app with a 3D map or provide a way to toggle from a 2D map to a 3D map, such as a button labeled 3D next to the zoom navigation. A good default 3D perspective is north facing and greater than 60 degrees. If applicable, use a 3D **marker** to indicate the user's current location. Points of interest can be labeled with perspective markers, so-called billboards, or **callouts**. Always try to declutter the view by reducing the size of items that are farther away or removing them completely.

Clicking in the map should identify and highlight the first thing that is "hit," in the same way as an arrow that is shot at a target. Two ways to visualize the selected item are (a) additive highlighting that changes the identified object's color so that it stands out from the rest and (b) subtractive highlighting that dims the environment around the object. **Location finder** should add a 3D marker, and **search** should highlight the feature, such as a building or office, through either additive or subtractive highlighting. Consider adjusting the perspective so that the view of the highlighted feature isn't obstructed.

Be aware of the following pitfalls when working with 3D maps:

- **Sea of walls:** 3D objects such as buildings, landscapes, or extruded objects can conceal other objects, and indoor spaces are especially prone to that. Try to simplify walls, remove them if possible, or show schematic walls that are only knee-high so users can still see "inside."
- **Easy to get lost:** Users can easily lose orientation as they zoom in and change angle and perspective. Add a north arrow button to reset the perspective, and add signage to help users get oriented. Consider disabling orientation or angle change to avoid this situation.
- **Right perspective:** Seeing an object from the right angle often involves additional navigation to find the right perspective. Consider generating **placemarks** that include predefined camera settings.
- **Unfamiliar navigation mechanism:** 3D maps should follow common keyboard shortcuts and mouse controls. It's also helpful to explain navigation controls to the user either up front or through a help menu. Remember that editing and placing objects in 3D are difficult tasks.

EXAMPLE

Kemayoran is an international trade center in Jakarta, the capital of Indonesia. It is the center of activity for international-scale commercial, service, and residential business. The Jakarta 3D Urban Regeneration–Kemayoran app (figure 3.16) uses a **3D** map to visualize projects, plans, and indicators within Kemayoran. Users can **search** for projects, apply **attribute filters**, and select a project to show its details. The details view follows the **focus on the subject** pattern to switch from the overview to the project. In this view, to highlight the buildings that belong to that project, their façades are rendered using detailed 3D models, whereas the other buildings are stylized in simple gray shapes. An **info panel** describes the project in more detail. Users can navigate back to the main view of the whole area by clicking the back arrow on the top.

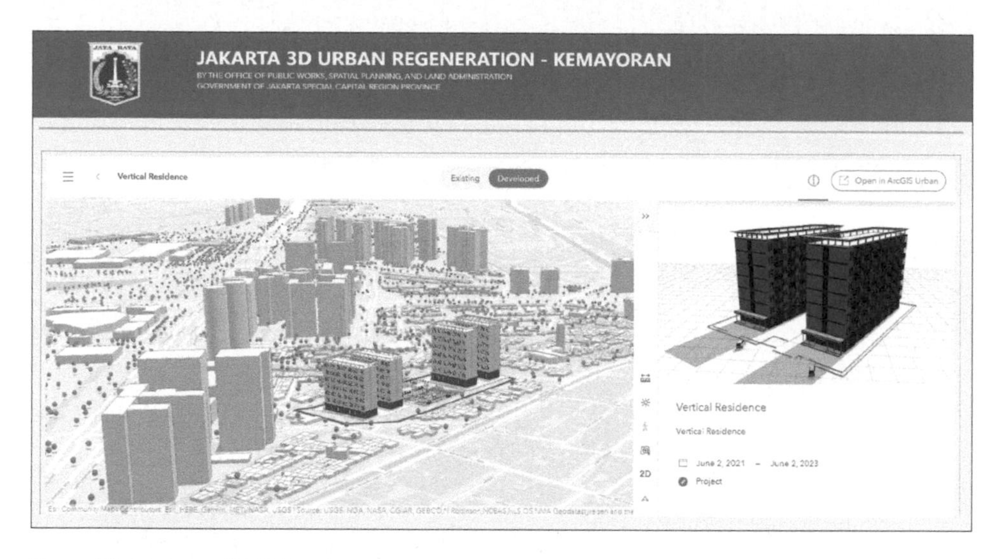

Figure 3.16. Jakarta 3D Urban Regeneration app lets users explore urban planning projects in the Kemayoran area of the city. Source: Dinas Cipta Karya, Tata Ruang dan Pertanahan Provinsi DKI Jakarta and Esri Indonesia.

References

Esri, ArcGIS API for JavaScript documentation, "MapView," https://developers.arcgis.com/javascript/latest/api-reference/esri-views-MapView.html.

Esri, ArcGIS Indoors, Space Planner, https://doc.arcgis.com/en/indoors/space-planner/introduction-to-space-planner.htm.

Esri, ArcGIS Platform, GeoEnrichment, www.esri.com/en-us/arcgis/products/arcgis-platform/services/geoenrichment.

Chapter 4

Dealing with complex data

You may have heard the phrases "Using this app is like drinking from a firehose" or "I feel like I'm drowning in a sea of data." These analogies express a feeling of being overwhelmed and disoriented, which is usually a bad sign. Why is this happening? How can you help your users make sense of the excess of information available and focus on their task instead? This chapter discusses solution strategies and patterns that will help you deal with the inevitable data overload in rich web apps.

Reducing the amount of data

Adding data to an interactive map is straightforward. The difficult part is to add data in a valuable way without impeding the UX. This is especially true for large datasets, sometimes also called high-density data or large data, that inherently come with the following challenges:

- **Performance:** Users are impatient and don't want to wait for large datasets to be loaded; the app becomes sluggish and less responsive when handling high-density data.
- **Bandwidth:** Many people, especially on mobile devices, have slow or no internet connection.
- **Visual clutter:** It's difficult to find the needle in the haystack of layers that display lots of overlapping features.

The most obvious warning sign is when your user stares at a crowded map, feels overwhelmed, and doesn't know where to start. Another reason a user might stare at the screen inactively is because data continues to trickle in, and they don't know whether important information has finished loading yet. One immediate remedy is to reduce the amount of data so it is as simple and clear as possible. Simplify complex features or thin overlapping features on the server side as part of your data preparation.

Remember that the goal is to make it as simple as possible by showing only the information needed and to include and display only what is relevant for the purpose of the app. This applies to the number of layers and the number of features within each layer.

Solution strategies

What complicates the display of complex data is that the density is relative to the map scale, and as builders, you have little control over your user's navigation habits. When, for instance, the points representing accidents are nicely dispersed at large scales, they become denser as the user zooms out to smaller scales. In situations such as this, various strategies exist to improve the UX (figure 4.1), as follows:

- **Filtering:** Remove unwanted features through filtering by attributes, spatial extent, time, or selection.
- **Layer visibility:** Toggle layer visibility explicitly or based on scale range.
- **Interactivity:** Allow user-triggered animations to highlight, dim, or show features.
- **Dynamic markers:** Dynamically group or cluster features.
- **Data-driven visualization:** Use meaningful symbols to show spatial patterns, trends, outliers, and anomalies.
- **Aggregation:** Generalize through grouping multiple objects with similar characteristics.

The choice on which strategies to apply depends greatly on user needs and business needs. When a Florida resident cares only about accidents in their state, at smaller scales you may choose to cluster the accidents or hide the accident layer altogether. For a first responder, on the other hand, it would be hazardous to remove accidents, so either an explicit way to filter the data or clustering may be the way to go. Insurance brokers will most likely be interested in statistics such as total insured value of the vehicle, so aggregation by county or state is probably the best choice here. Additionally, business needs may require accidents to be grouped by assignment to a portfolio, which means that the UI must then provide an attribute filter for portfolios. As you can see by looking at the latter example, understanding the structure, hierarchy, and relationships of the layers as well as the features with their attributes will directly influence the UI elements you can provide. We often observe that interface design drives data architecture and therefore shouldn't be an afterthought.

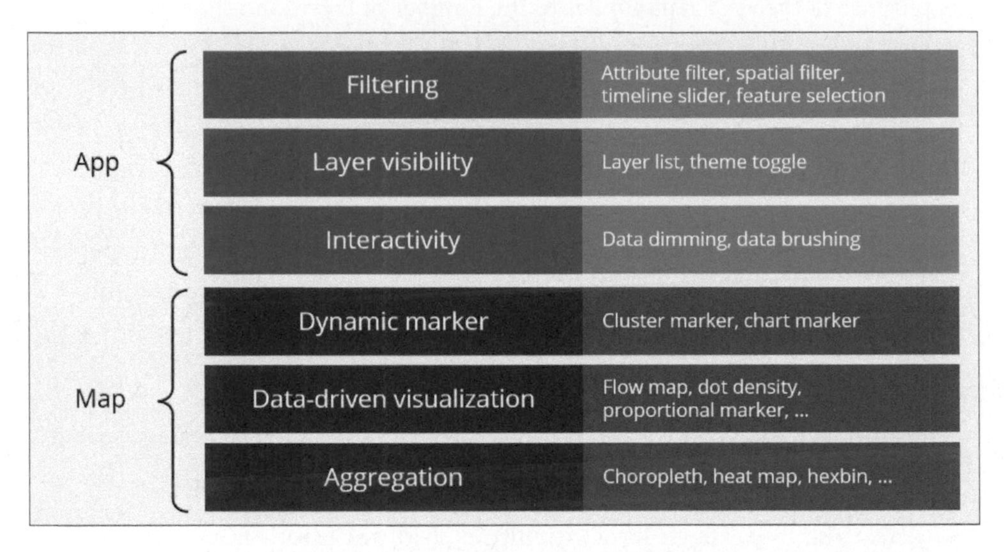

Figure 4.1. Solution strategies for dealing with complex data through the app and the map.

Most of these strategies go beyond classic mapping solutions as you know them from static maps. They all require interface elements that structure, chunk, or manipulate the data that's visible in the map and should embody the following characteristics:

- **Be reactive:** Respond quickly to user input
- **Be iterative:** Easy to refine
- **Be explicit:** User-initiated actions and user control
- **Be contextual:** Related to current extent and scale

Remember that the goal of a map app is to learn or do something. It's the builder's job to understand what that something is and to provide a path for how to get from the big picture to the details. Give users a UI that allows them to explore their problem and view the data from different angles.

Patterns

The patterns in this chapter align with the solution strategies described earlier. The first set of patterns addresses the different ways to apply filters through the app:

1. Attribute filter
2. Spatial filter
3. Timeline slider
4. Feature selection
5. Buffer (www.mapuipatterns.com/buffer)

The second set of patterns introduces you to dynamic markers and how they can help you simplify the data through the map. This simplification usually involves some form of highlighting or visual transformation into pieces and includes the following patterns:

6. Cluster marker
7. Chart marker

User-triggered interactive animations to highlight features or patterns in features are still less common in map apps but can be useful to improve the UX, as follows:

8. Data dimming (www.mapuipatterns.com/data-dimming)

Adjusting layer visibility either explicitly or through other means is the GIS mechanism that probably has been around the longest. The following patterns related to layer visibility can be studied online:

9. Theme toggle (www.mapuipatterns.com/theme-toggle)
10. Layer list (www.mapuipatterns.com/layer-list)

Aggregation patterns are useful for reducing the number of points visible at smaller scales while simultaneously showing patterns in your data. These patterns are as follows:

11. Choropleth (www.mapuipatterns.com/choropleth-map)

12. **Heat map** (www.mapuipatterns.com/heatmap)
13. **Legend** (www.mapuipatterns.com/legend)

Data-driven visualizations are powerful ways to look at data. They are also well documented elsewhere and therefore mostly discussed online:

14. **Flow map** (www.mapuipatterns.com/flowmap)

Attribute filter

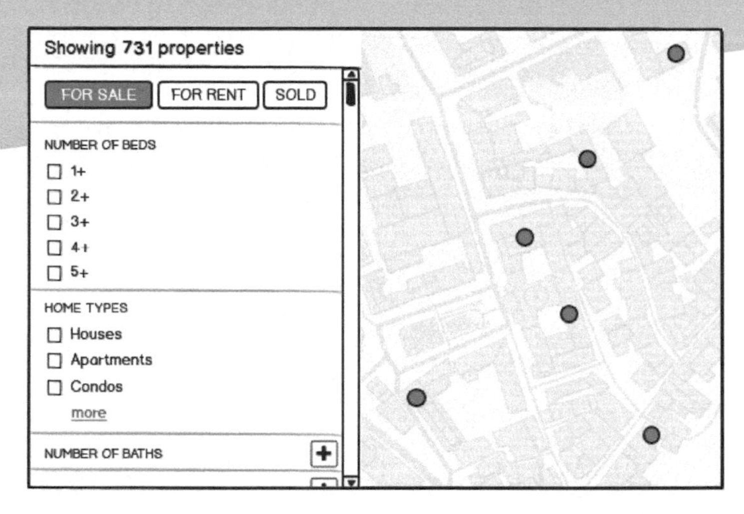

Figure 4.2. Attribute filter pattern.

WHAT

The attribute filter pattern (figure 4.2) reduces large datasets into more meaningful and manageable selections based on attribute criteria.

WHY

Filters are omnipresent in modern apps and, like search, are the most important mechanism to find the data that best suits a user's needs. Search aims at finding a particular place, object, or group of objects that have a common value. Attribute filter, on the other hand, works by refining and reducing the number of items from a full list of data. Seeing all the data can be overwhelming and complex, appear crowded, and isn't necessarily useful because users are typically interested in only a specific subset of the dataset. The goal of the attribute filter is therefore to help users reduce the data to the smallest number of options that are relevant to their task. This goal aligns with Hick's law, which describes the time it takes for a person to make a decision as a result of the possible choices. Increasing the number of choices will increase the decision time logarithmically. Conversely, reducing the number of choices will aid people in making better decisions more quickly. The attribute filter helps users make better and faster choices.

WHEN

Filtering is a common way to reduce the number of features displayed on the map or in a list. It allows users to progressively remove what they are not interested in and so leaves them with what they are interested in. Common use cases include consumer apps in the real estate, travel, and shopping space and enterprise apps in the insurance and agriculture industries.

HOW

Attribute filter comes in many flavors and can vary greatly in complexity and length. Sometimes a simple drop-down list suffices while other times the filter is long and warrants its own panel. On mobile devices, filters are often a series of buttons and drop-down choices just below the header. You also may use other components such as the legend, charts, or timeline slider as the filter control.

Start by showing all the features and provide a form-based interface that allows users to filter based on attribute values. This approach is sometimes referred to as "data first, filter later." Construct the attribute filters so that users get to their desired selection as quickly as possible. Although the practical construction of your filter will vary from case to case, the following rules provide general guidance:

- Bring important filters to the forefront.
- Carefully select the attributes to be filtered. Domain values tend to work well.
- Use check boxes to opt in and option buttons, also called radio buttons, to toggle between mutually exclusive options.
- Use radio buttons when few options exist; use a drop-down list when options extend beyond a handful.
- Consider adding counts to groups of options to indicate the impact of a filter choice.
- Consider graying out impossible filter options or remove them.
- Range sliders may return zero results, so consider restricting minimum/maximum values to indicate the available attribute ranges.

- Provide smart default selections, such as the most common choices, all options selected, or none selected.

The best practice for applying filtering logic is to construct a filter query that filters "OR" between attribute groups and that filters "AND" between the options of a group. Filtering works best when changes to the filter return asynchronous results and display immediate feedback on the number of results, such as "Showing 731 properties." Make it clear that a filter was applied and is active by adding small hints or explicit text. Provide an easy way to reset and clear the filters.

Finally, don't forget to consider the **empty state**. Some filter combinations may result in zero results to be displayed on the map or list. Provide feedback on what happened and an easy way to reset the filters.

EXAMPLE

The Census Business Builder: Small Business Edition (CBB: SBE, figure 4.3) app provides potential and new business owners access to a range of information, such as demographic, socioeconomic, workforce, consumer spending, or employment data. For instance, if a prospective restaurant owner wants to research the best place to open a new, high-end, specialty dining option, they would add "restaurant" to the cluster of selected industries. Then they would select a primary mapping variable such as "total population" and optionally add a second variable to enable bivariate mapping. After the choropleth map is rendered by the selected geography level (such as by zip codes), the user can apply an **attribute filter** to refine the visualization. This future restaurant owner may want to add a filter that shows only zip codes in which "average income is greater than $100,000." Applying this filter will gray out all the zip codes that don't match the filter criteria. The remaining geographies appear highlighted and indicate better target locations for opening the new restaurant.

Figure 4.3. The Census Business Builder: Small Business Edition app provides easy access to information about potential customers and similar businesses for those starting or planning a business.

Spatial filter

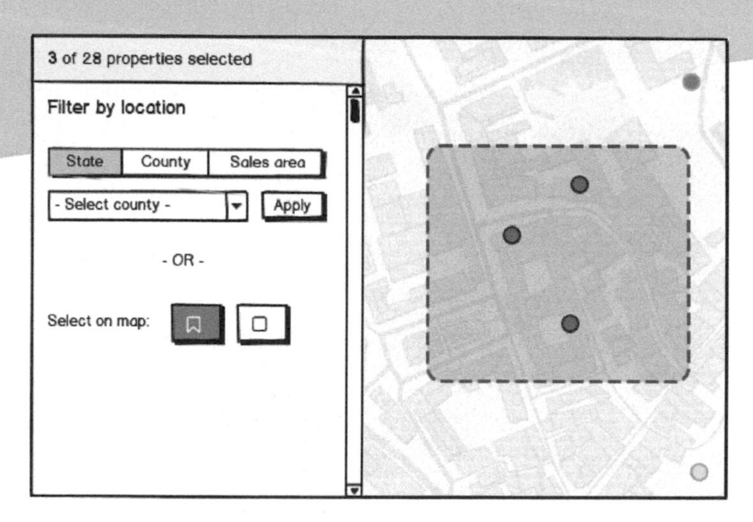

Figure 4.4. Spatial filter pattern.

WHAT

The spatial filter pattern (figure 4.4) reduces available features by selecting a specific geographic region or area of interest.

WHY

Location, location, location is not only the mantra in real estate but also true for most consumer apps in which location is what people care about the most. This location can be a place or any well-defined geographic area such as country, state, district, city, service area, or postal code. The spatial filter is sometimes also referred to as filter by geography and provides an explicit method of specifying the desired geography for further exploration or analysis. In natural language, a spatial filter would say: "Show me only all the blah-blah that are within geography x,y,z. Remove all the others."

WHEN

Reducing the results by geography is an option widely used in enterprise apps, analysis apps, and dashboards. Dashboards, for instance, use the spatial selection to update the map display and any other component that is connected to the data, such as charts, KPIs, or tables. Spatial filters are less common in consumer apps such as real estate or travel apps, where the **location finder** is more prevalent. Location finders search for houses or hotels within a city but aren't filtering the dataset—they only modify the extent to zoom in on the destination without removing the other options around them.

HOW

Show all the features without any geographic restrictions. Let users define a geography through interface components such as a drop-down list. Order the list of options alphabetically and consider adding a search field to reduce the options if the list is long. When the user selects an option from the list, all the features that do not fall within the selected geography will be removed.

The spatial filter usually filters features by an attribute of the feature set. Alternatively, the filter may run a server-side or client-side query. Client-side queries have the advantage that the data is already available in the browser and the filter can be applied immediately. Server-side queries often require an explicit button click to apply the filter query. Regardless of the filter mechanism, always provide an easy way to clear the filter.

For lesser-known geographies, help users by letting them browse a hierarchical set of geographies, such as selecting a state from a drop-down list to reveal a second drop-down list for selecting a county within that state. The following patterns describe alternative approaches on how to filter data spatially:

- **Browse geographies** lists all the options for easy browsing instead of lengthy drop-down lists.
- **Feature selection** describes how the map can be used to draw an arbitrary filter area.
- **Buffer** defines a spatial filter by creating a buffer zone around a selected feature.

EXAMPLE

The USDA National Water and Climate Center (NWCC), in Portland, Oregon, supports the Snow Survey and Water Supply Forecasting Program and Soil Climate Analysis Network (SCAN) pilot program. As part of the USDA's Natural Resources Conservation Service, the NWCC is responsible for producing and disseminating accurate and reliable water supply forecasts and other climatic data to its wide variety of users. The Water Supply Outlook app (figure 4.5) allows users to access water supply reports in an interactive way. There are two ways to apply spatial filters to focus on the area of interest. The first spatial filter is a drop-down list of available states in the top-right corner. Selecting a state, with Colorado selected in the figure, will remove the other states from the map, zoom the map to the state's extent, and update the content panel on the left. As part of this update, the list of basins within the selected state changes also. This list becomes a secondary spatial filter. Once the user clicks a basin, the map will filter out all the other basins and zoom to the selected one on the map. The other content components, such as the gauge charts, will update accordingly. Spatial filters are the main mechanism to drill down to the desired area.

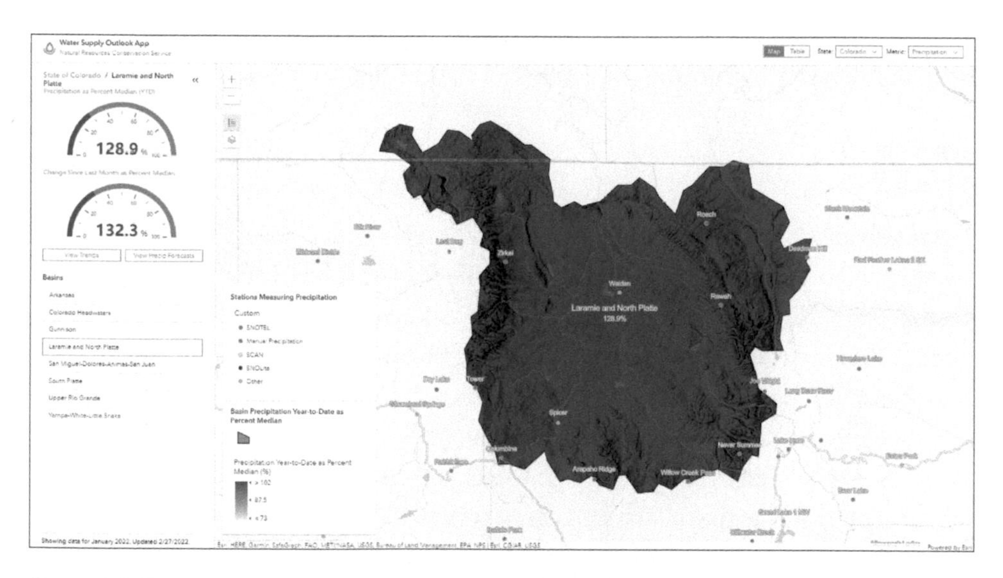

Figure 4.5. The National Water and Climate Center's Water Supply Outlook app offers an interactive way to access water supply reports. Source: US Department of Agriculture, Natural Resources Conservation Service.

Timeline slider

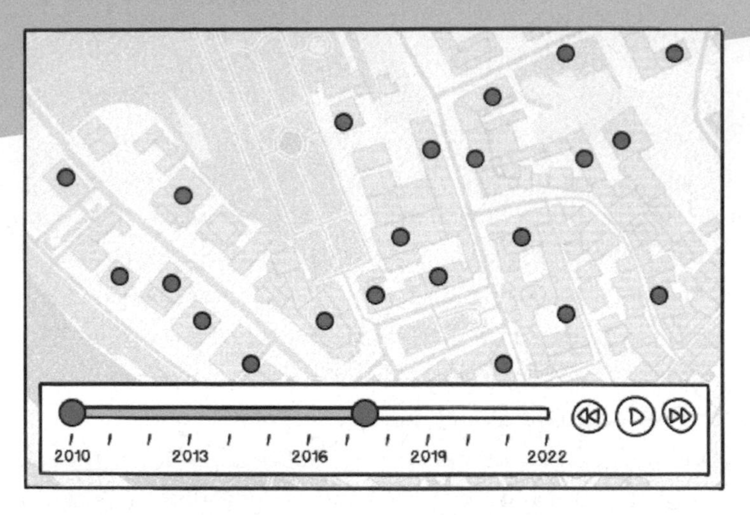

Figure 4.6. Timeline slider pattern.

WHAT

The **timeline slider** pattern (figure 4.6) uses the slider to control the features that correspond to a certain time span. Timeline slider reduces the amount of data to display by showing only the data within that time span.

WHY

Observations or measurements are rarely single occurrences. Phenomena happen continuously and change their values over time. Knowing when something happened can be as important as knowing where it happened. Data that is being collected for the same feature but at different times is referred to as temporal data. In most cases, the location of the feature is stationary, such as a weather station, streetlight, property, or geographic boundary. At other times, the location changes also. The data is then referred to as spatial temporal or spatiotemporal, which is important for tracking and analyzing assets, movements, or shipments across geographic areas and over time.

WHEN

The time span can be fixed or cumulative. Fixed time spans typically use hours, days, months, or years and show totals within that span for comparison. As the timeline slider advances, only features that fall into the applicable time span are visible. Cumulative time spans show progression from the beginning to the current time extent. Examples for apps using fixed time spans are crime mapping by day of the week or sea level rise by year. Apps that use cumulative temporal data, for instance, include sales dashboards to show growth or health apps to show virus outbreaks.

HOW

Provide a horizontal slider component with two knobs or levers to control the selected time span. Use a single knob instead of the levers to represent a fixed unit, such as a day or year. Add labels to clearly state the start and end of the complete time extent and tick marks at periodic intervals in between. The size of the timeline slider should be relative to the time extent and the available screen size. Remember that sliders are difficult to manipulate and should be used to select time approximations. If picking exact values is important, choose other input components, such as date pickers, drop-down lists, or button bars. Consider snapping the slider to human-friendly values, such as hours or days.

Another common interaction method for the timeline slider is to define a fixed time span, such as a day, week, or year, and use that duration to "step" forward and backward in time using playback controls. That way, you are guaranteed to see all your data in order and with equal time slices.

The following three approaches can be used to see cumulative time—for instance, stores opened since the year 2000:

- Default the start date to the year 2000 and show only one adjustable knob to change the end date.
- Allow users to toggle from the one-knob cumulative display to a fixed time span that is controlled by two knobs.
- Provide a lock button to temporarily freeze either the start or the end knob.

In the map, filtered features should be hidden by default, but if their presence is important, you may choose to apply opacity instead so they are still visible in the map. Alternatively, consider adding a "firefly" effect that gives points a bright, glowing appearance. This effect works best in dark, desaturated basemap layers and is a nice way to show growth or expansion by how "the map lights up."

EXAMPLE

Crime Time (figure 4.7) is a sample app in Esri's COOL MAPS series. The app visualizes San Francisco crime data by neighborhood, day of the week, and hour of the day. It makes good use of the container that holds the timeline slider by also showing the title, a drop-down list to select the crime type, the legend, and a visually interesting way to select the day of the week. The timeline slider has a single green lever and tick marks with labels indicating the time of day. It is set to play by default with a small button to pause. The app uses cluster marker symbols to show the cumulative total number of crimes as indicated with a number inside the black inner circle. The light-green circle around the inner circle grows proportionately to the number of crimes. Additionally, the cluster marker adds another circle in dark green to show the total incidents within the currently selected hour.

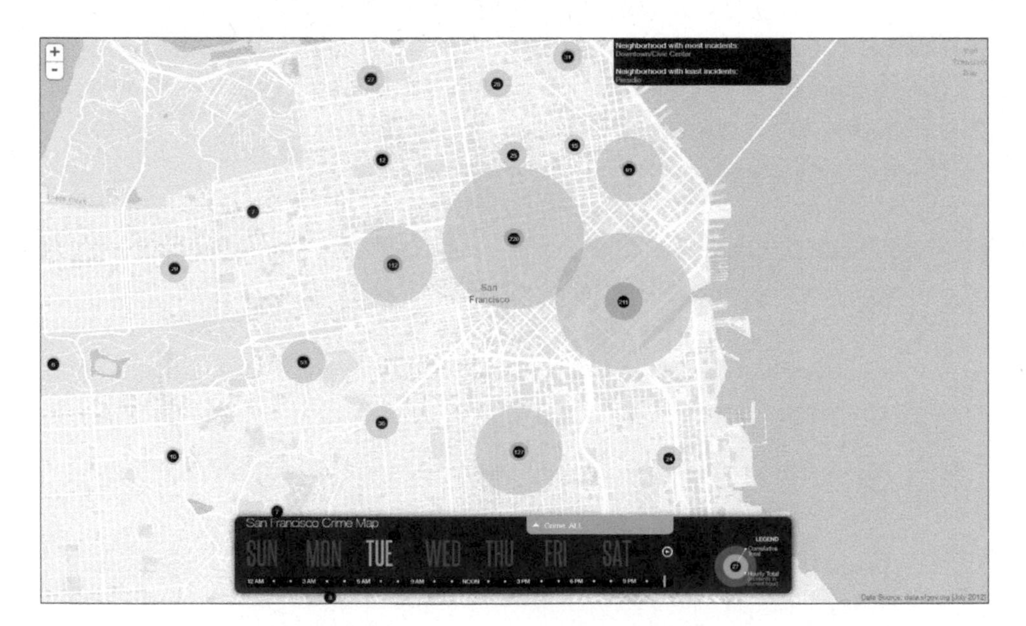

Figure 4.7. Crime Time app uses a timeline slider to show cumulative totals and crimes per hour.

Feature selection

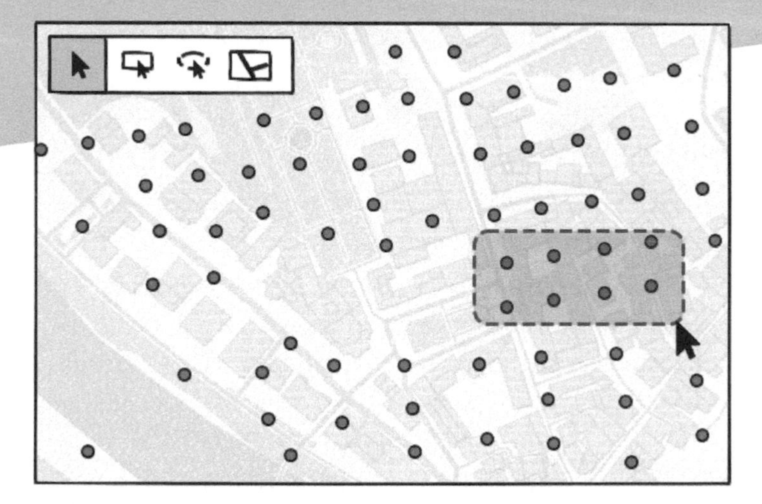

Figure 4.8. Feature selection pattern.

WHAT

The feature selection pattern (figure 4.8) uses selected features on the map to create a subset from all the available features. It helps shift the focus from the whole collection of features to the ones that are relevant.

WHY

Selecting items is a common input mechanism to gain further insights into the associated data. The goal is to create a new collection of features that can be used as input for further analysis. The feature selection is like the spatial filter, in which the filter criterion is a user's selected area on the map. The difference is that the selection isn't bound to a discrete, predefined area but based on a custom-drawn shape, such as a rectangle, circle, polyline, or freehand drawing.

WHEN

The output of the feature selection becomes the input to another action, such as analytic calculations, editing, sharing, or export. Adding features to a selection set can also be used to update one of its attributes across all selected features at once or to create focused reports that include only the selected features. Other common uses include calculations of aggregated property values for a planned construction site or display of average insured value for an area affected by natural hazards.

HOW

Feature selection requires a temporary change in how users interact with the map. This change must be triggered by the user and moves the map from its default behavior into selection mode. Users can trigger selection mode by pressing a single button or selecting from a series of buttons, also called a toolbar or button bar, with mutually exclusive selection options such as rectangle, circle, or freehand. The following list describes the different selection methods.

- **Single selection:** This is the expected default selection and turns an unselected feature into a selected feature, and vice versa.
- **Selection by rectangle or circle:** Drag the shape to select one or many features.
- **Freehand selection:** Click and draw a line to sketch an area that encloses features inside, sometimes referred to as lasso selection.
- **Polyline selection:** Click to start drawing, continue clicking to create lines, and double-click to connect the last point with the first point. This selection method is difficult to use for non-GIS professionals, so use freehand selection instead.

Once the user has entered selection mode, disable the toolbar, show a button to end the selection process, and optionally add descriptive text that provides additional guidance. Guidance can be descriptive text next to the button or a ToolTip that follows the pointer. Consider providing feedback on the count of currently selected features. This count information is helpful especially when features were selected outside the currently visible extent. Selections should always be additive; the exception to this rule is single selection. An effective method to display the individual features that are part of the selection is a grid table or **location list**. Once the user ends the selection process, provide an option to clear or start over.

Selected markers need to change their appearance to visually distinguish them from unselected features of the same kind. Use a highlight color such as yellow or cyan or alternative symbology without altering the meaning of the symbology in place.

Another way to add individual items to the selection is by providing an action button inside the **info pop-up** or the list item of a grid table. For instance, an app to report streetlight outages could show streetlight information inside the pop-up paired with the option to add this item to the list of broken streetlights.

EXAMPLE

This sample app for national parks (figure 4.9) demonstrates how to select features on a map. The selection set can be used to inspect their attributes or export them to another app. To activate the tool, the user clicks the select features by rectangle button and draws a rectangle on the map. Once the rectangle is drawn, a query returns the features within the rectangle for display in the grid table in the image. Features that are selected remain unaffected visually, whereas features that are outside the rectangle have a blur effect applied.

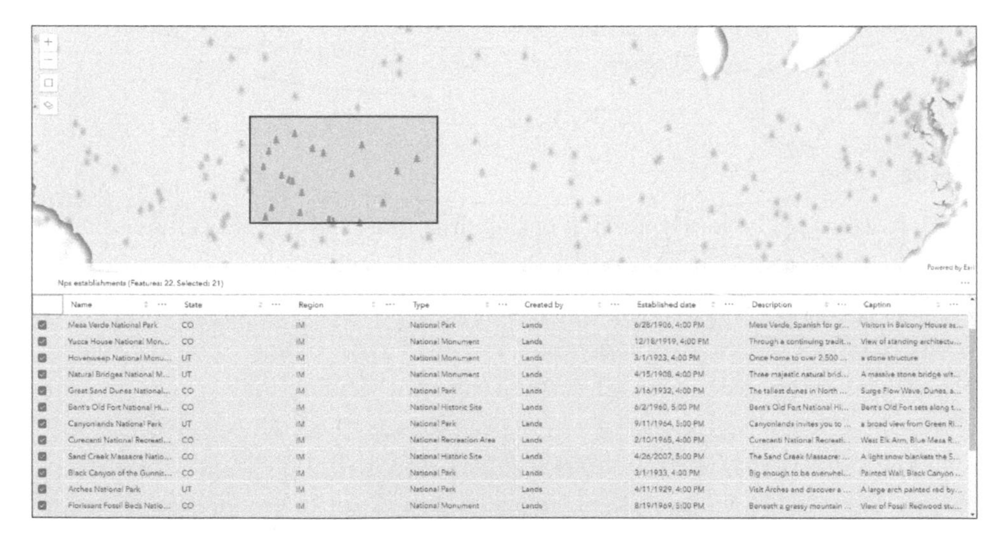

Figure 4.9. User selects national parks in the map to inspect their attribute values in a table.

Cluster marker

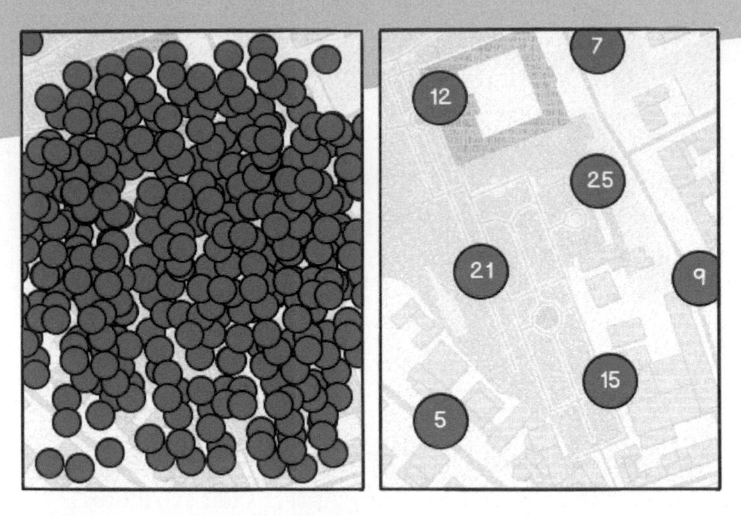

Figure 4.10. Cluster marker pattern.

WHAT

The **cluster marker** pattern (figure 4.10) is a common method to avoid visual clutter by grouping points together, and many apps that display point data benefit from this pattern. Cluster marker differentiates overlapping features that are difficult to distinguish and interact with.

WHY

Data layers can have any number of point features, and it is almost inevitable that features that are nicely dispersed at large scales become denser and start to overlap as the user zooms out to smaller scales. Ways to avoid this congestion are to reduce the number of markers, use smaller symbols, or limit the zoom level, but often these approaches aren't the right solution for a variety of reasons. Clustering the markers into groups is useful for cutting down on the visual clutter and informing users that large volumes of points were grouped together. These groups, or bubbles, are created based on screen distance from one another. Larger bubbles signify more points, and as the user zooms in, the bubbles become smaller until they are no longer necessary and the individual points become visible.

WHEN

There are two types of cluster markers: the original type that labels the cluster by the number of features it represents and the less common type that exposes aggregated metrics, such as the sum or average of an attribute for the clustered features. Apps that represent the former type usually focus on actual locations, such as events, incidents, or policies. Apps of the latter type are often analytic in nature and allow, for instance, insurance providers to see annual average loss of their policies or real estate agents to see average prices of recent home sales.

HOW

Group individual features that are close to each other into a single point symbol. This is called a cluster and resembles a bubble. Use symbology that is like the original point marker but larger in size. Often the diameter of the cluster marker increases proportionally to the number of markers it represents, indicating its magnitude. Cluster markers are usually circular with a number inside so remember to ensure sufficient contrast between the label and the circle. All cluster markers must have the same color, even if the individual features that make up the cluster are visualized with different colors based on an attribute.

As the user zooms in, the number of clusters decreases, and individual markers start to appear instead. Zooming out has the opposite effect, and markers are consolidated into cluster markers again. Consider adding animations that show how points are being added to or removed from the clusters. If the user is zoomed into the lowest zoom level and markers are still clustered, maybe because the points are on top of each other, you may want to provide a way to explode or "spider" the cluster marker where leader lines connect the center with the individual points arranged in a circle around it.

Hovering the pointer over a cluster marker symbol will outline the area—the bounds—used to consolidate the features. Clicking a cluster marker symbol should zoom to the bounds of the cluster. Alternatively, an **info pop-up** with details or summary statistics on that cluster can be displayed.

Different algorithms exist to calculate clusters. The most common algorithm calculates counts based on distance from each other, which results in a balanced distribution. Predefined grids work well, too, if the cluster marker is positioned based on the average relative distances of the markers within each cell. Avoid using geographic boundaries such as states to create the clusters. If you want to visualize counts by geography, use **choropleth** map visualization instead.

EXAMPLE

What does "home" mean to you? This question is deeply emotional and has a different meaning for each one of us. It is almost always connected to a very special place, and that's exactly what the Out of Eden Walk HomeStories app (figure 4.11) is trying to highlight. Thousands of people from all over the world have used this app to upload a photo together with a short description of the place that feels like home to them. Over time, the app has become a kaleidoscope of life stories. Every story is represented by a dot on the map. Showing individual dots works well at large scales, but at smaller scales, the points overlap quickly. To avoid this undesired behavior, the app uses cluster marker to aggregate overlapping points into bubbles that indicate the number of points within. When a user clicks one of the cluster markers, the app assumes that their intent was to see stories for that region and zooms to that area. In many cases, individual points start showing up, or else a second click on another cluster is needed to zoom in even further. The list of stories in the left panel updates to show only relevant stories visible in the current extent (see **extent-driven content** in chapter 6, "Building Single-Purpose Apps"), and clicking a story card toggles to the details of that story (see **list and details**, also in chapter 6).

Figure 4.11. The Out of Eden Walk HomeStories app asks people to submit multimedia stories, with images and narrative text, describing a place that feels like home to them and why.

Chart marker

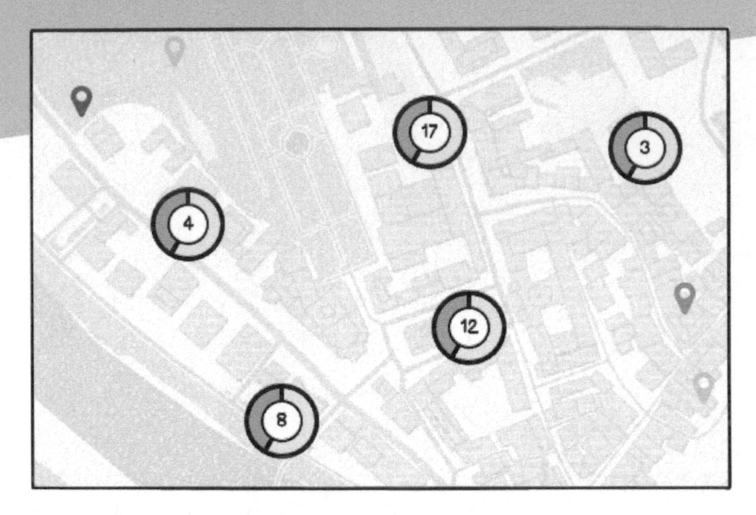

Figure 4.12. Chart marker pattern.

WHAT

The chart marker pattern (figure 4.12) uses charts to enrich aggregated data or to visualize a second attribute of clustered data. Adding tiny charts to the map is a powerful visualization technique to communicate additional aspects of the data.

WHY

It's easy to symbolize a single observation through color, shape, or size, but as soon as multiple observations are grouped into clusters or aggregated by geography, you are limited to showing values by statistics, such as count, sum, or average. Potentially important insights into the data get lost. For instance, seeing that an area has lots of schools is good to know, but having a pie chart showing how many of those schools are public versus private is even better. Useful chart types include trend lines to indicate change over time, histograms to illustrate distributions, or pie charts to show proportions.

WHEN

The chart marker pattern adds insights to the map beyond the already chosen visualization. You may have chosen cluster markers to group traffic incidents, but only a pie chart marker can show the proportional breakdown of incidents by type.

Another use of chart marker is adding a second value to the data. This allows for displaying multiple topics at the same time instead of mapping a single attribute alone. For instance, show the maximum occupancy of a hotel as a number in the center and the current occupancy as a doughnut chart around it. In that way, it's easy to see that a hotel has 100 beds and is 90 percent booked.

HOW

Add tiny charts to the map that visually look like markers. Pie, doughnut, and line charts tend to work well since they can be shown in a simple and condensed way without axes or coordinates. Doughnut charts have the additional benefit that you can add a number to the center to show the totals. If the chart markers are based on **cluster marker**, the charts will update as the user pans or zooms the map according to the cluster algorithm in use.

Add hover effects to show additional information about the chart, such as data for each slice of a pie. Click events can be used to invoke an **info pop-up** that shows additional details, such as the breakdown of the individual features that make up the chart.

Follow general best practices for chart creation. Pie or doughnut charts, for instance, should start their slices at "12 o'clock" and use few categories and well-distinguishable color palettes. As with **choropleth** maps, displaying a **legend** is important to understand the meaning of the chart.

If the charts show data that belongs to a fixed geometry, such as a county or zip code, consider using scale dependency to disable the chart markers at smaller scales to avoid overlap or switch to larger geometries as input for the charts.

EXAMPLE

The CovidPulse app (figure 4.13) takes a novel approach to visualizing reported coronavirus disease 2019 (COVID-19) cases and deaths. Although most COVID-19 dashboards show daily snapshots of new case counts with the all-too-familiar red graduated circles, CovidPulse uses chart markers in the form of sparklines to visualize trends over time and across counties. *Sparkline* is a term coined by Edward Tufte, who described them as "data-intense, design-simple, word-sized graphics" to show the variation in measurements—in this case, variation in COVID-19 infection counts. The map shows negative trend lines in red and positive trend lines in green. **MapTips** reveal additional indicators, and an **info panel** docked at the bottom provides more detailed charts. As the user zooms out to the national level, the app starts to use state data as the input data to avoid crowded and overlapping charts.

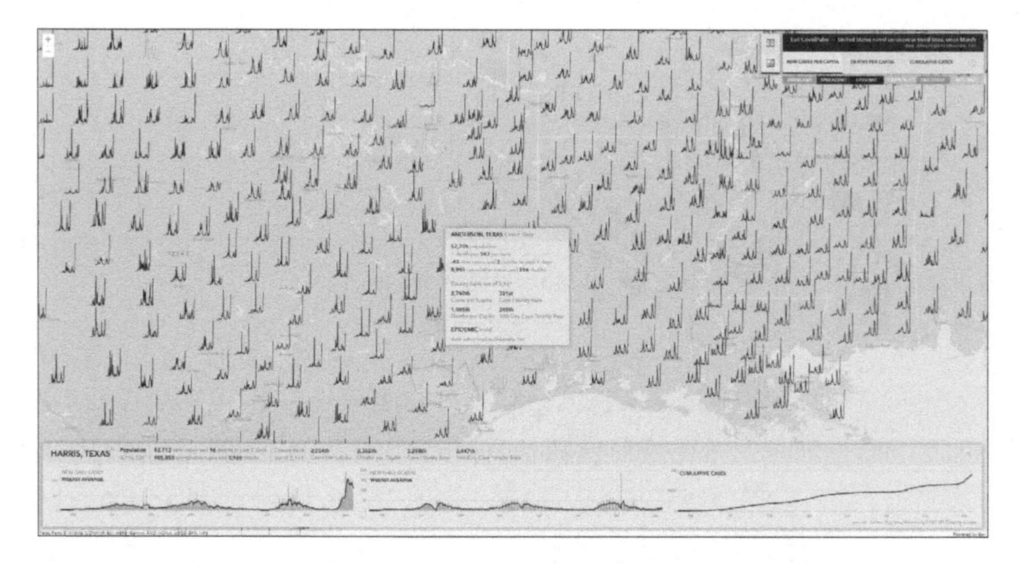

Figure 4.13. CovidPulse visualizes COVID-19 trends over time.

References

"Data-Driven Visualization," https://developers.arcgis.com/documentation/mapping-apis-and-services/visualization/data-driven-styles.

"Hick's Law," https://en.wikipedia.org/wiki/Hick%27s_law.

"High Density Data," https://developers.arcgis.com/javascript/latest/visualization/high-density-data.

Nelson, John, and Jinnan Zhang. 2020. "COVID Pulse." *ArcGIS Blog*, August 31, 2020. www.esri.com/arcgis-blog/products/arcgis-living-atlas/health/covidpulse.

Chapter 5
Designing for mobile devices

Mobile devices are ubiquitous. People who rely on small, portable, interactive, connected, and location-aware devices to retrieve quick, relevant, and specific answers have outnumbered users of large-screen desktop computers—not to mention all the people who need precise information and directions in real time and on the go. With advances in device technology and broadband speeds, the need for mobile design is going to grow even more. This chapter will help you understand the differences between responsive and native apps, learn about the challenges and opportunities of mobile design, and guide you in providing usable workflows and mobile patterns.

Responsive versus native apps

The term *mobile apps* is used interchangeably for responsive web apps and native mobile apps. Although each option comes with its own distinct advantages and disadvantages, the workflows and patterns discussed in this chapter apply to both types, and their differences are due mostly to access to native features, device capabilities, and cost considerations.

Responsive web apps use cascading style sheets (CSS) media queries to accommodate different screen sizes and adjust content. Historically, responsive design has been an afterthought, but more recently developers have adopted a "mobile first" mentality that optimizes the experiences with small screens in mind first before scaling up to larger screens, such as tablets and desktop. Responsive apps are usually cheaper and faster to develop, provide cross-platform accessibility, and are indexable and shareable. On the flip side, they require an internet connection and have only limited access to native device features and sensors.

Native mobile apps, or simply native apps, are developed and tested for a specific mobile operating system (OS) such as Google Android or Apple iOS. To run them on a mobile device, you must first download and install them from their respective app store. If you want to target multiple platforms, you must program the same app for each OS independently. Native apps leverage native phone features, such as accelerometer, Global Positioning System (GPS), or camera; work offline; and often provide better performance and nicer designs with personalized preferences. Disadvantages include the need to download and install yet another app on your phone and that building apps for multiple environments adds development costs for each supported OS.

A third, hybrid approach exists that marries the best (and worst) of both worlds. Developers program and test the app in a single language and build it to multiple native target platforms. ArcGIS AppStudio is a great example of a suite of productivity tools for pursuing this hybrid approach.

AppStudio gives you everything you need to build geoenabled cross-platform native apps from a single code base. Learn more at www.esri.com/en-us/arcgis/products/arcgis-appstudio.

Challenges and opportunities

"Hey, I know this place, I've been lost here before." What sounds like a joke is the core value of mobile map apps—to successfully navigate through unknown territory and ever-changing environments. Discovering nearby places and finding and exploring locations are key tasks, and apps that embrace these needs can reap big rewards with their users. However, mobile design comes with a series of challenges that, if not taken into consideration properly, can turn the opportunities into pitfalls. Following is a list of challenges:

- **Technical limitations:** Low-bandwidth networks and areas without connectivity interrupt interactions.
- **Screen size:** A wide variety of small to tiny screen sizes leads to visual clutter and information overload.
- **Proficiency:** Low proficiency in reading and handling maps requires extra effort to guide users and provide easy answers.
- **Input controls:** Difficult touch interactions result in unexpected behavior, sometimes also referred to as "fat-finger syndrome."

- **Situational context:** Attention span and focus are low; distractions are omnipresent and difficult to predict.
- **Environmental influences:** Changes in external influences such as bright daylight, rain, indoor settings, or even wearing gloves affect user behavior and the ability to use apps.

Responsive design means more than simply adjusting the size and arrangement of elements. It extends to taking care of content density and gesture events and providing adequate touch areas. Data visualizations and interactions created for desktop devices cannot be assumed to work on mobile devices. For instance, hover doesn't exist on mobile devices, so any mouse click events must be backed up by touch events. Always strive to follow platform guidelines such as *Apple Human Interface Guidelines* or Android documentation for app developers, and use familiar UI elements in your app.

Beyond platform guidelines and considering these challenges, you must also consider the following best practices for mobile maps:

- Mobile maps must be simple to comprehend.
- Choose a basemap that gives as much reference as needed for the task but doesn't visually overwhelm or distract.
- Simplify, generalize, or cluster features to find a well-balanced, scale-to-marker ratio that minimizes overlapping markers while meeting touch target dimensions.
- Increase the size of the markers and action buttons to meet current touch target standards of at least 44 × 44 pixels.
- Remove or hide controls that are not essential or less needed.
- Maximize the space available for the map by removing unnecessary headers, footers, and panels.
- Minimize the use of toolbars and other buttons that float on top of the map.
- Use device sensors to locate, position, direct, and assist users with their tasks.

How to use a mobile map

Now that you understand the challenges and opportunities of mobile maps, consider different scenarios in which you can use a map in a mobile app. The most common use cases for using a map in a mobile app are the following:

- **Location finder:** Users must find a nearby location that meets their criteria; their task is to find the details of a specific result and get directions to it.
- **Exploration:** Users explore a specific topic such as restaurants by browsing an area of interest in the map; different from using **location finder**, they are usually looking for inspiration and general ideas.
- **Collection:** Users that are on the scene must complete a task that includes finding the physical location of an asset and collecting or editing data about that asset.
- **Navigation:** Users need to reach their destination, which requires an app to provide routing and turn-by-turn **route directions**.

Whereas many apps strive to optimize a single use case, more comprehensive apps such as the Ohio Trails app DE**TOUR** (figure 5.1) offer two or more uses. DE**TOUR** users, for instance, can use the app to explore and browse any area within the state of Ohio to get ideas for their next hiking trip.

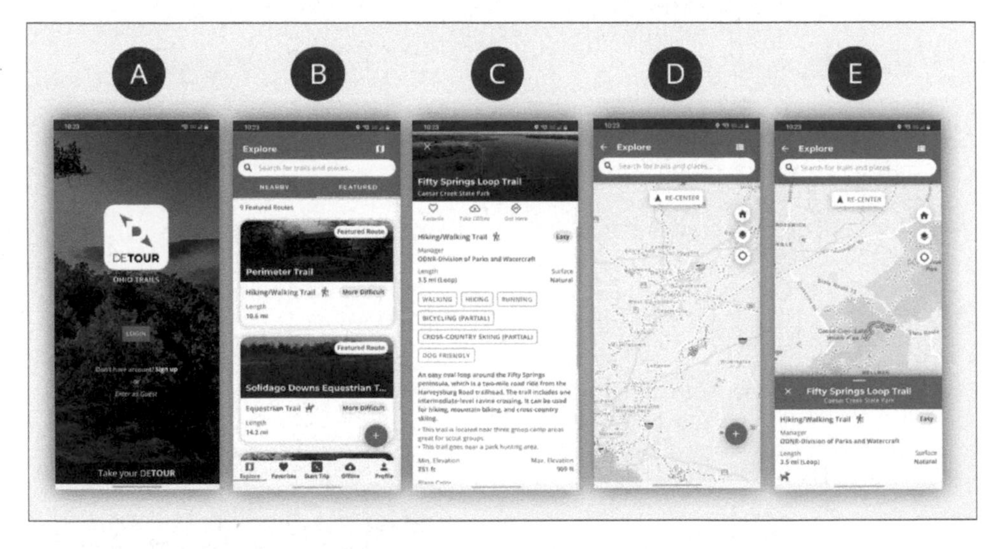

Figure 5.1. Individual pages of the mobile workflow for creating your own hiking path as shown in the Ohio Trails app DE**TOUR**.

The app also offers an easy way to find trails following the location finder workflow (figure 5.2 and described in more detail below). Once the desired trail has been selected, the start trip workflow provides navigation along that specific trail, which includes showing points of interest and allows trail users to submit issues about the trail. The app uses a global navigation menu at the bottom of the screen to toggle between these sections besides other menu items such as favorites, offline maps, and user profile.

The location finder workflow consists of a series of pages (shown in figure 5.2 with letters *A* through *F*). Deviations from this workflow are common but vary mostly in layout, content, and arrangement of functionality. The following is a summary of the pages and their functions:

A. The landing page (*A*) is the first step in a user's journey. It's the place to collect important information about the user, such as account credentials, and often includes search and locate me. Knowledge about the area of interest such as a user's current location, home, travel destination, or another points of interest will be used on subsequent pages. Exploration apps may omit this page and jump right to the list or map page.

B. The list page (*B*) lists places near the area of interest. After specifying the location, a location list is the expected next step. Each item in the list typically includes distance from location, useful information such as business hours or services provided, and action buttons to see more details (*C*) and to get route directions (*F*). Selecting an item from the list should open its details. Repeat the search on top of the page and populate it with the previously selected location information. Consider adding an attribute filter to reduce the number of results. Provide an easy way to toggle to the map (*D*), either through a floating button at the bottom or through a button next to the search. Don't forget to consider the empty state in case no results are available.

C. The details page (*C*) provides details on a selected item. Show useful information about the item beyond those already included in the list item or info pop-up. Optionally, you may want to include a link to navigate to the map (*D*) and include a reference map showing the item's location. Refer to the list and details pattern for considerations about toggling between list (*B*) and details (*C*).

D. The map page (*D*) provides the spatial context. Use **full map** layout pattern to show as much map as possible and avoid **swipe** ambiguity that comes with **partial map** or **embedded map** layouts. Exploration and navigation apps are less focused on lists and may direct users straight to the map. Emphasize the **search** and spend additional time on cartography and **marker** design. Clicking a marker will open the pop-up page (*E*). Consider **cluster markers** or other techniques described in chapter 4, "Dealing with Complex Data," to simplify the map. Add a button near the bottom of the screen that toggles to the list page (*B*). Exploration apps may list locations as floating cards in the map. This list is usually placed near the bottom of the page and can be scrolled horizontally.

E. The pop-up page (*E*) shows either an **info pop-up** or an **info panel**. The pop-up content should be consistent with the list content described in (*B*). Add action buttons to open the details page (*C*) and the directions page (*F*). Consider docking the pop-up to the bottom of the screen rather than floating it next to the marker.

F. The directions page (*F*) provides **route directions** and can be embedded into the app or link to an external routing app. Receiving step-by-step directions is probably the single most important reason why mobile users work with maps on mobile devices, and that's why the list page (*B*), details page (*C*), and pop-up page (*E*) all link here. Always prepopulate the destination and, if possible, the starting point.

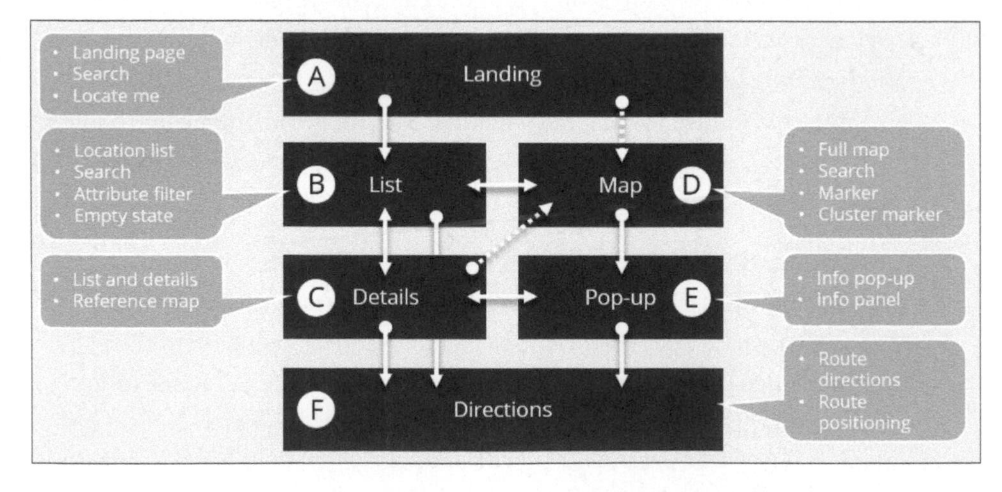

Figure 5.2. Schematic of a workflow diagram for a mobile map app.

Patterns

The first set of patterns in this chapter addresses the single most important aspect of designing for mobile users: location. Users need to find their current location, see and share this location in the map, and navigate in a 3D space. The patterns are as follows:

1. Locate me
2. Blue dot
3. Floor selector

Bandwidth is another important characteristic of mobile apps. Number 4 empowers the user to download data only when they need it, and number 5 describes workflows that anticipate low or no connectivity:

4. Search this area
5. Offline maps

The last pattern in this chapter deals with the emerging field of augmented reality:

6. Augmented reality (AR)

You can find the following mobile map app patterns online:

7. Share live location (www.mapuipatterns.com/share-live-location)
8. Nearby (www.mapuipatterns.com/nearby)

Locate me

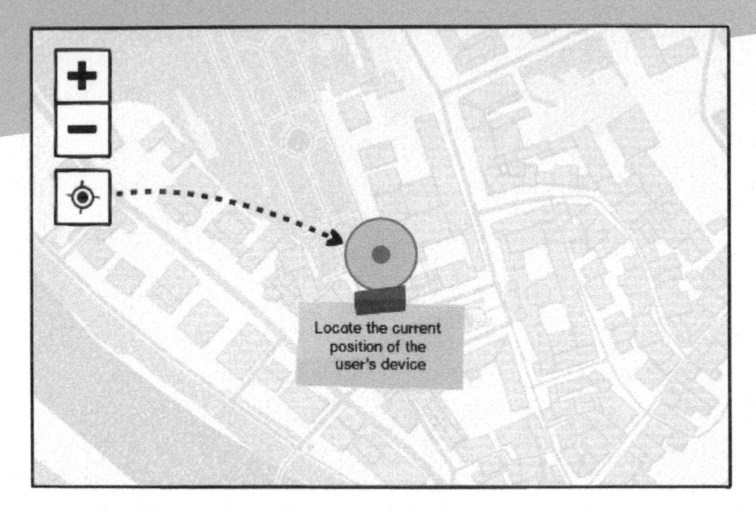

Figure 5.3. Locate me pattern.

WHAT

The **locate me** pattern (figure 5.3) helps users quickly find their current location in the map. It is a crucial mechanism for providing meaningful context about the user's location and surroundings.

WHY

Users on mobile devices are focused on their immediate area and tend to have a narrow center of interest or activity. They are usually on a small-screen device and may be in a rush with little time to handle complicated data input or map navigation. Having a simple way to identify and locate a user's real-world geographic location saves them time and enhances their experience with the app.

WHEN

Locate me is a common mechanism on mobile devices to find the user's current position. Sometimes referred to as geolocation, the estimated geographic position is determined via mechanisms such as GPS, Indoor Positioning Systems (IPS), or services that use network routing addresses. Mobile apps use locate me for route finding or directing people to nearby places such as tourist attractions or available parking spots. Store locator apps use locate me to identify nearby stores or locations.

HOW

Provide a button that triggers a process to determine the latitude and longitude coordinates of the user's geographic position. Label the button Locate Me or Near Me, or simply show the locate me icon that looks like a target symbol. Place the button in a corner of the map—a good place on mobile devices is near the bottom where users can easily access it with their thumbs.

After invoking the locate me action, zoom and center that map at the user's current position and show a blue dot. Zoom to a map scale that shows enough context and minimizes the need for further navigation. The actual extent will depend on the task and the data—for instance, a parcel finder may zoom straight to the parcel at a very large scale, whereas a store locator probably wants to include some stores around that location and therefore decides to zoom to a medium scale.

Although the app is centered on the user's location, the Locate Me button may be hidden if it doesn't affect other buttons shifting into its vacant space. Once the user starts panning the map and their current location isn't centered anymore, the Locate Me button reappears so users can re-center on their location again. Alternatively, the icon may change from showing a target icon with a dot in the center while centered to using the same icon without a dot otherwise.

As an alternative to the Locate Me button, a link may be used. This approach is especially useful when more verbose labels such as Find a Hospital Near Me or Local Deals Tonight are needed. Place the link near a location finder or search and consider adding the Locate Me icon adornment for clarity.

Users should always understand why the app is asking for their location. Provide context for its purpose and, if possible, hint at how this information will be used.

EXAMPLE

Dubai Electricity and Water Authority (DEWA) developed the solar calculator (figure 5.4), a mobile responsive web app that encourages citizens of the United Arab Emirates to adopt a more sustainable way of generating electricity and reduce their carbon footprint. The app offers two ways to find a location. One way involves typing an address, place, parcel, or makani, the official geographic addressing system in Dubai, into the location finder. The other way is to click the Locate Me button. The button is positioned in the bottom right corner of the interface so that the thumb can reach it easily. After clicking it, the browser may ask for permission to access the device's location and then zoom the map to the user's current location. A blue dot shows the identified location. Because the focus of this app is to visualize solar potential, the zoom level is large enough to display building roofs. The different shades of yellow indicate suitability for solar panels on the specific roof surfaces. Tapping on a roof reveals additional information about the solar potential.

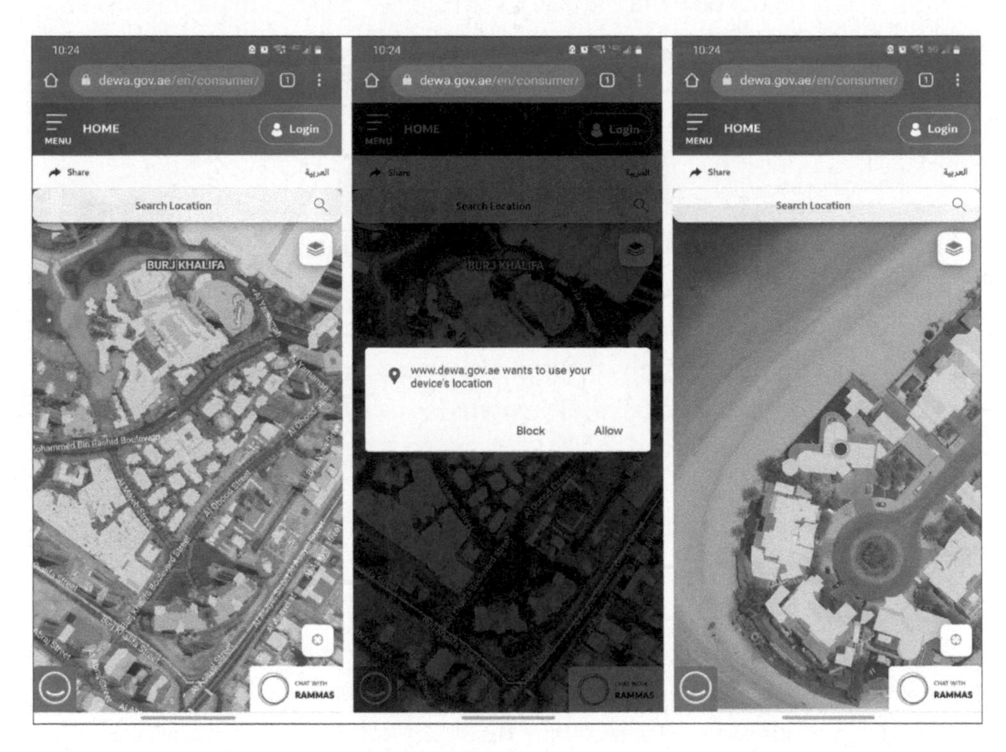

Figure 5.4. Dubai Electricity and Water Authority's solar calculator allows citizens to assess a building's solar suitability. Source: Dubai Electricity and Water Authority (DEWA).

Blue dot

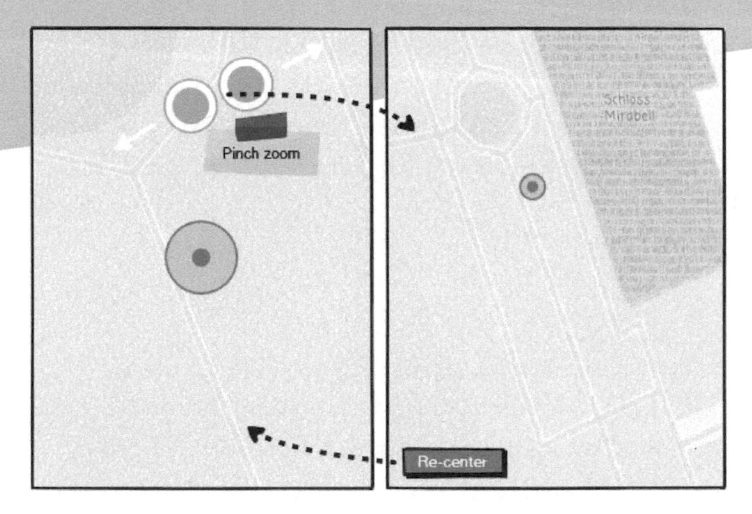

Figure 5.5. Blue dot pattern.

WHAT

The **blue dot** pattern (figure 5.5) allows mobile users to see their current location in a map to understand their surroundings. Blue dot is specific to mobile apps where movement and change in position of the user's device is expected. Tracking this movement means that the user can focus on their task without expending energy to constantly pan and zoom the map to their current location.

WHY

People who navigate indoor and outdoor spaces rely on their location to make decisions on features around them. From their point of view, the world gravitates around them—they are the center of the scene with important objects surrounding them. Because mobile users are constantly moving, their expectation is that the map "follows" them. This experience differs from the usual map navigation behavior that requires the user to provide explicit pan or zoom instructions to change the extent.

WHEN

Blue dot is extremely powerful for navigating along a route so that, for instance, a driver can focus on the street while the app follows the car to show relevant surrounding features such as traffic hazards, gas stations, or upcoming turns. Other common use cases include locator apps such as **store locators** and apps that support people on the scene such as hiking apps and field data collection apps.

HOW

Blue dot is an app mode that needs to be triggered by the user, usually by pressing the **locate me** button or a button to start navigation. The map zooms and centers on the user's current location and places a blue dot at the center of the map. This indicates the location identified through GPS, IPS, mobile phone mast triangulation, internet protocol (IP) location, built-in phone sensors, or a combination of these.

As the user moves the mobile device, the map moves accordingly while keeping the blue dot centered at the preconfigured scale. Under certain conditions—for instance, approaching an important intersection where the user needs to see the full scene—the scale may temporarily change. Once this condition has passed, the app should zoom back to the default scale.

If users interpose themselves in the navigation experience by zooming or panning the map, the map should stop following the users' movement. The app will instead show a button labeled Re-center, usually placed in the bottom-left corner, which allows users to return to the tracking mode.

Two methods to finish the navigation experience should be provided: cancel and done. A user should be able to cancel and back out of navigation by clicking a return button, usually indicated by a left-pointing arrow in Western languages. The second method is a notification trigger once the destination has been reached. This notification may be a modal overlay informing the user of the fact that they have arrived, often paired with additional information about the location and further actions that can be taken.

Last, any positioning is only as accurate as the mechanism available and used to locate the device. Indicating this accuracy is a useful feedback mechanism so that users can make decisions accordingly. A nice way of presenting this feedback is through the size of the circle. Often a larger blue shade around the dot means a higher margin of positional error. Some apps display an indicator cone to show the direction a user is facing. The narrowness of the cone correlates with how well the device is calibrated for direction.

EXAMPLE

The ArcGIS Indoors mobile app (figure 5.6) uses floor-aware indoor maps to help executives, supervisors, mobile staff, and employees better understand, maintain, and operate workplace environments. A user's location is the key to finding rooms, tracking assets, and routing to facilities. The app uses search, locate me, info pop-up, and floor selector to explore the 3D indoor space. The blue dot is an important pattern to indicate the user's current location. It is always active but can be outside the visible map extent, in which case the user can tap the Locate Me button to re-center the map on the blue dot. Indoors allows employees without a permanent office location to book a room before they visit. When they arrive at the workplace, the app routes them to the booked room so they can check in and use it.

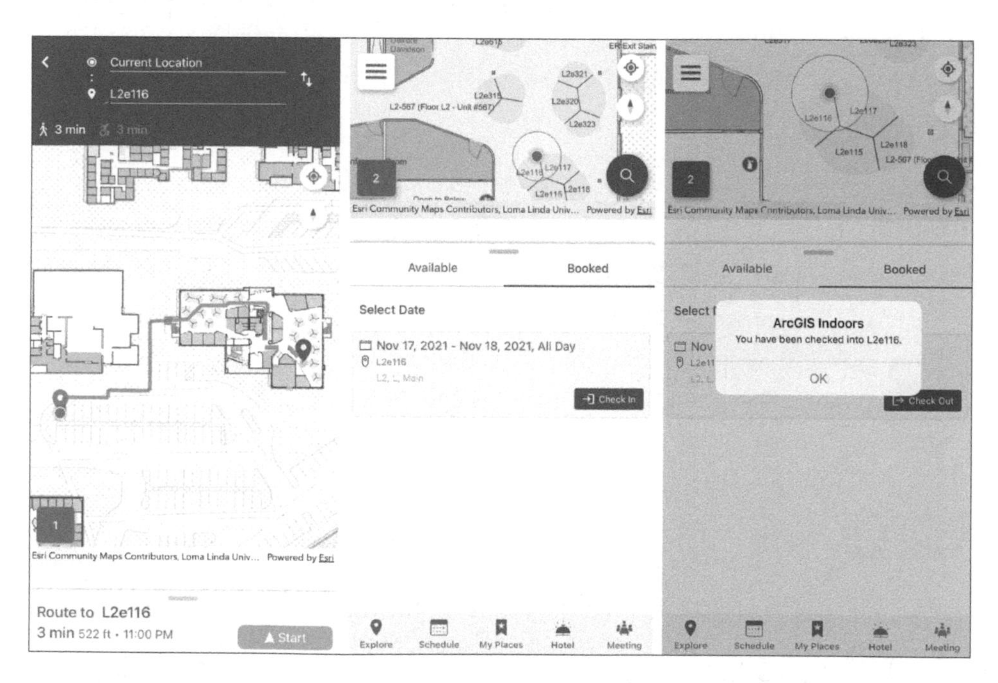

Figure 5.6. ArcGIS Indoors uses the blue dot to indicate the user's location inside indoor spaces.

Floor selector

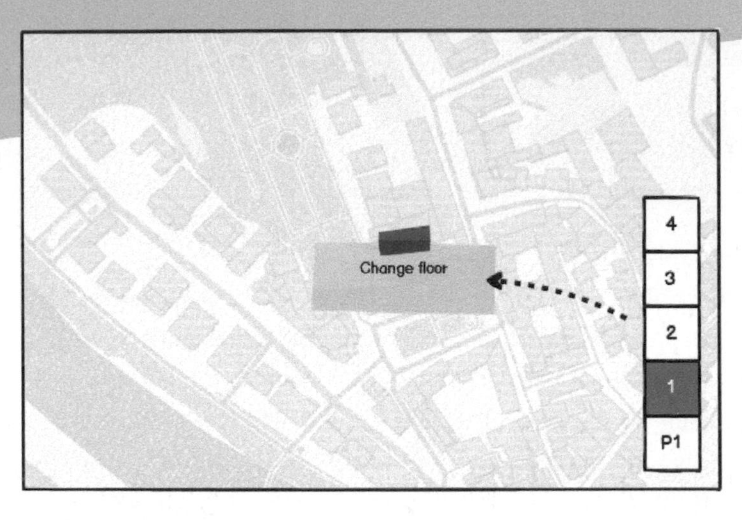

Figure 5.7. Floor selector pattern.

WHAT

The floor selector pattern (figure 5.7) lets the user switch between floors or levels in a 2D indoor map. The floor selector, sometimes also referred to as level selector, floor picker, or floor filter, becomes an essential UI control providing a means to toggle between floors and gives visual feedback on the currently selected floor.

WHY

Indoor maps are inherently 3D in that they often span across multiple floors or levels. When displaying these maps in a planar 2D map, only the topmost floor plan can be visible. If all floor layers were transparent, the display would be impossible to understand. There is clearly a need to select a single floor at a time with the ability to toggle to a different floor level.

WHEN

Any app with a floor-aware map needs a way to select the level or floor. This is especially true for 2D maps in which higher floors visually overlap the floors beneath. Show the floor selector as soon as an indoor view is active, even on mobile devices with limited space. Typical use cases for the floor selector include apps that cover event venues, transportation hubs, corporate or university campuses, or hospitals. What these apps have in common is the need to help visitors navigate, orient employees, track assets, or plan safety and security.

Sometimes regular "outdoor" maps turn into "indoor" maps. This switch can happen explicitly or implicitly. Explicitly means a user needs to click a building and select it from a list. Implicitly means the app detects the building automatically as the user zooms or pans close enough to its extent.

HOW

The floor selector consists of a series of stacked buttons that have labels to indicate the level. Labels are usually floor numbers (1, 2, 3, and so on) or abbreviated level names such as L1 for level 1 or GL for ground level. Using longer labels is acceptable if they don't cover important map content. An alternative to buttons is a drop-down list, which consumes less space and allows for longer labels. The disadvantage is that options aren't immediately visible, and an additional click is needed to select a floor.

Place the floor selector in a corner of the map, often at the bottom of the screen to mimic the vertical order of a building. Position it on the opposite side of map tools such as **zoom control** or **locate me**. Select the ground floor by default unless it was triggered through **search** or a query that requires the selection of a specific floor. Keep the height of the component at a fixed maximum that doesn't stretch across the full height of the map, and restrict the number of buttons to a manageable size such as five to not overwhelm the user with unnecessary options. Navigating to surplus levels should be done through a scroll bar without the need for up and down arrows above and below the floor buttons.

The two methods for enabling the floor selector are as follows:

- **On demand:** The floor selector isn't visible until the user selects a building. Alternatively, the app determines that the user has zoomed in close enough to switch on the floor plans for that building. The floor selector turns off again when the user clicks an exit button or zooms away from the building. This approach aligns well with the user's mental model, in which people enter a building, use the elevator to change floors, and then exit the building again.

- **Always on:** The floor selector is visible as soon as a floor-aware map is detected. The ground-level floor plan is then selected by default. Always on works best for apps with only one building.

If your data model includes multiple buildings or spans across multiple sites, consider adding functionality to select the site and buildings within that site. You may also indicate the hierarchical nature of the site/building/floor relationship using breadcrumbs, a list of links representing the current location and its ancestors, usually separated by the > symbol.

EXAMPLE

The ArcGIS Indoors app (figure 5.8) uses a floor selector to switch between floors of a selected building or facility. To enable the floor selector, the user first selects a building to see inside. By default, the floor plan of the first floor becomes visible. Toggling to another floor reveals the offices on that floor. For sites with more than one building, the floor selector has a button with a building icon that allows users to search for other buildings. If multiple location sites are configured, users can follow the same workflow to select another location from the hierarchy of sites. Once the user is done visiting a building, they can click the Close button on the floor selector to leave the building, hide the floor plans, and minimize the floor selector.

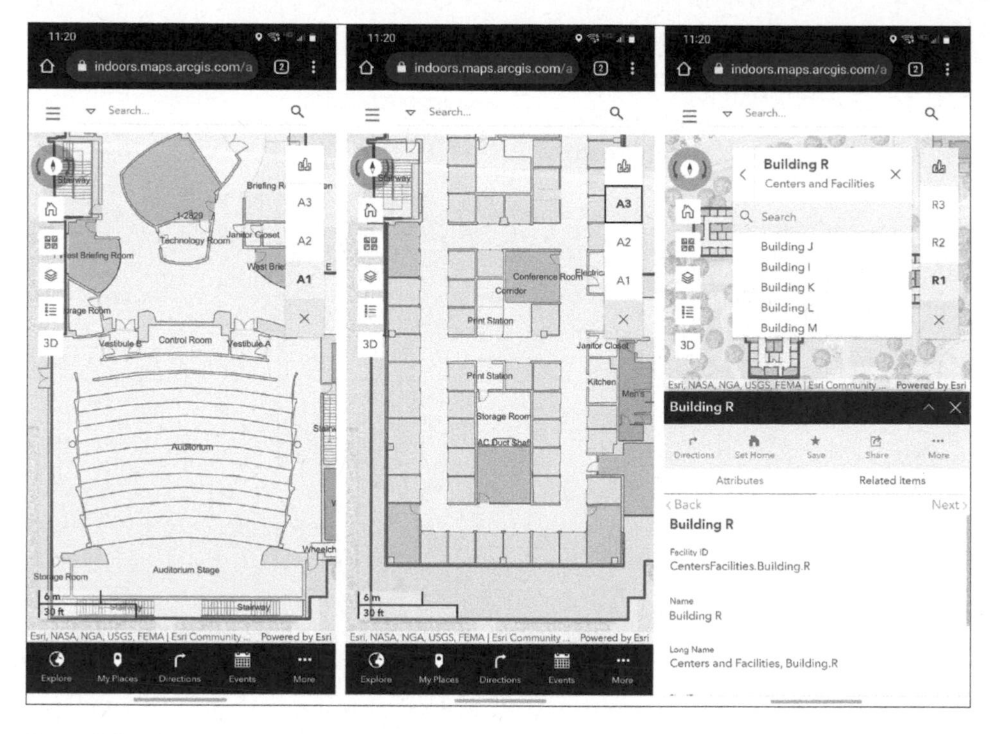

Figure 5.8. Indoors uses the floor selector to switch between levels of a multistory building.

Search this area

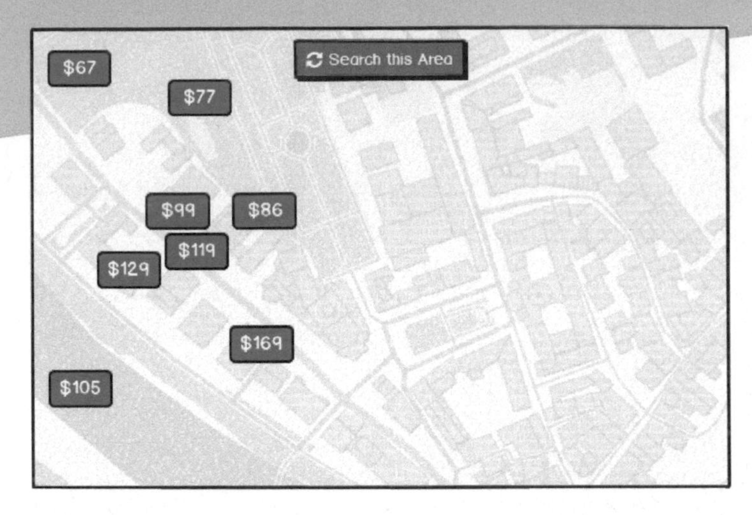

Figure 5.9. Search this area pattern.

WHAT

The **search this area** pattern (figure 5.9) allows the user to search various points of an area, giving the user control over when to update features in the map. Search this area avoids updating features in the map until the user instructs the app to do so. It can avoid confusion and frustration for the user.

WHY

Any time a user pans or zooms the map, the visible area changes and new features must be loaded from the server. Besides initial features being removed from sight, this can lead to the problem that users may be left wondering whether data exists in this area or hasn't loaded yet. Unless the features are already available on the client browser, each small pan event requires a server request to retrieve the new features, which consumes unnecessary bandwidth on mobile devices. Another undesired side effect can occur when the map extent is tied to a list, which will change with every map move. This behavior is often expected but in other cases can lead to confusion or even frustration.

WHEN

The search this area pattern is often found in mobile exploration apps in which screen space is limited and data connections are unreliable. Constant panning and zooming requires features to be added or removed constantly. As a result, users lose track of features they have already looked at and become disoriented when new features trickle in slowly or not at all. This uncertainty can even lead to a feeling of helplessness and frustration.

HOW

The recommended implementation on mobile devices is to provide a button labeled Search This Area or Redo Search in Map. Place this button in the top or bottom center of the map or as part of the location list. Show the button whenever the map extent changes, and new features can be requested. Hide the button again after the user has clicked it to retrieve features for the current extent.

In situations in which bandwidth is less of an issue, your app can provide users with the choice to opt in to search the map as the extent changes. Instead of the button, add a check box labeled Search as Map Moves, Search as I Move the Map, Update Results When Map Moves, or something similar. Apply the following recommendations to the check box:

- **Checked:** Results will update automatically after each pan or zoom event. Keep the check box visible so users can opt out of automatic refreshes.
- **Unchecked:** Results will not be updated automatically. Instead, after a pan or zoom event has occurred, turn the check box into the Search This Area button. When this button is clicked, the app fetches the updated results and the button turns back into the check box.

The default state of the check box can be checked or unchecked and depends on its intended use. Consider using a small panel that floats in the top area of the map to host the check box and the button.

Because the main concern is bandwidth and data overload, apply additional safeguards to limit the number of features being fetched. One way is to implement a threshold that disables the Search This Area button after zooming out beyond a reasonable scale. Another way is to restrict the requested features to a smaller area, typically the size of a city—that is, as the user zooms out to a small scale, retrieve only features from a small area around the map center.

EXAMPLE

The mock-up in figure 5.10 exhibits an app that connects people with local businesses. Users can use their mobile app to find and explore these business sites in the map. The example shows a search for restaurants. As the user zooms or pans the map, new restaurant sites become visible and other sites may disappear. While the user tries to find their area of interest, the app avoids requesting new sites, helping to save bandwidth, and shows a Redo Search in This Area button instead. When the user clicks this button, the app removes existing points from the map and retrieves a new set of sites in the area. The new sites are added to the map as numbered **markers** so that the user can refer to them from the list that is accessible from the list icon in the header.

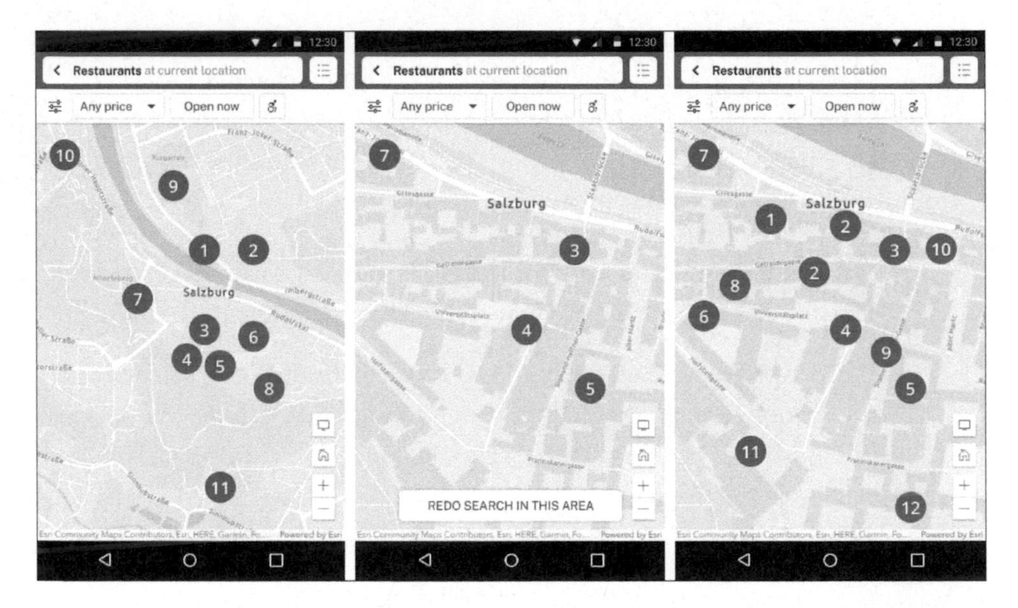

Figure 5.10. Mock-up mobile app for restaurants in Salzburg doesn't load new data until the user explicitly triggers a new search in the area.

Offline maps

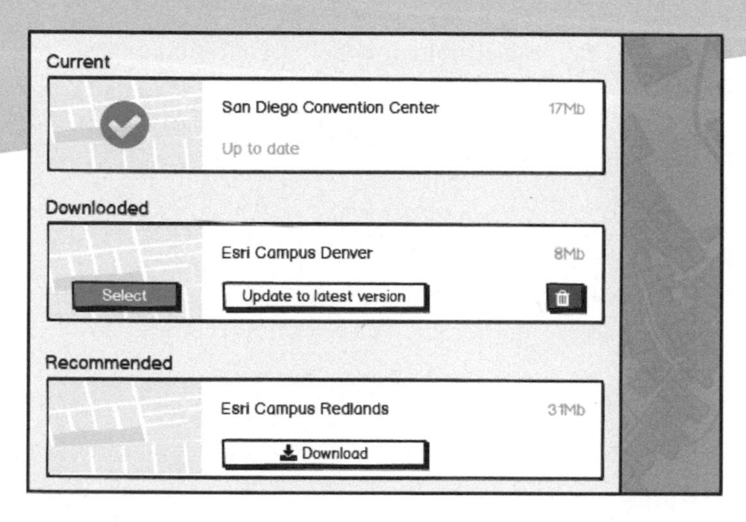

Figure 5.11. Offline maps pattern.

WHAT

The **offline maps** pattern (figure 5.11) uses offline maps that can be downloaded to a user's device to provide access to data when the need arises. They are useful when the user needs access to data in areas in which the internet is slow, mobile data is expensive, or the user cannot get online.

WHY

Offline maps, sometimes also referred to as offline areas, is most useful on mobile devices that are being carried to an area with data-serving limitations, such as slow, limited, or expensive internet connections. In anticipation of these limitations, users may want to download map content to a mobile device while still on a fast and reliable internet connection. This saves time and money and ensures that the content is available before traveling to the underserved area.

WHEN

Offline maps are important for mobile apps that include navigation and **route directions**, especially when traveling to rural areas with limited connectivity or poor reception. Hiking apps, for instance, must show reliable information about trails while the person is moving along unknown territory. Before the hiker travels into the outdoors, they can download trails and other infrequently changing data so it's readily available on arrival. Limitations to that approach are that live data feeds such as traffic information, transit updates, or alternative route suggestions will not be available without connectivity.

HOW

Provide a mechanism that allows users to download the map data for a specific region before encountering access limitations. The workflow typically includes the following steps:

1. Search for a place or navigate to an area of interest.
2. Start the workflow by clicking a button to Save Map Area.
3. Confirm the map area to be downloaded.
4. Specify the name of the data package.
5. Make sure the data package appears in the list of downloaded areas.
6. Allow users to update outdated or delete existing downloads.

Once an area is downloaded and the device identifies a slow or no internet connection, the app will use the local version instead of streaming live data. This so-called offline mode may also be triggered because the user has switched their phone to airplane mode. During offline mode, the app should add a message bar informing the user that data is temporarily or permanently unavailable.

The main considerations for the offline map experience are storage space and download size. Inform users of the anticipated package size before they start downloading it. If the size is too large, offer one of the following alternative ways to reduce its volume:

- **Select a smaller area.** For instance, include only Manhattan instead of the whole city of New York. Alternatively, consider offering predetermined and optimized map packages available for download in this area.

- **Adjust the data granularity.** Include only important data and omit unnecessary layers. The downside is that relevant information may be missing. Although it's okay to ask the user to confirm which data layers to include, a better approach is to let an algorithm decide instead of exposing the whole list.

EXAMPLE

The GreenbeltGo app (figure 5.12), developed by Greenbelt, Essex County's Land Trust, is a native mobile app that allows hikers to explore and navigate publicly accessible conservation lands in Essex County, Massachusetts. Knowing that cell phone range varies greatly in rural areas, the app offers users the ability to download offline maps for use on scene. A dedicated Offline Maps page lists the downloaded areas and allows users to start the process of saving a new map area. As part of this process, the user modifies the area to be saved by panning and zooming the map before clicking the Save Area button to confirm the selection. After the offline map is named, the map is downloaded and made available in the offline maps list.

Figure 5.12. GreenbeltGo allows hikers to download maps for offline use.

Augmented reality

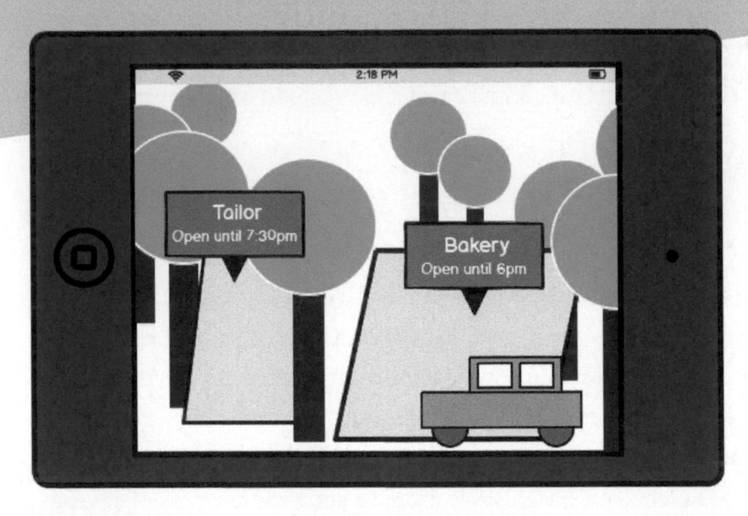

Figure 5.13. Augmented reality pattern.

WHAT

The **augmented reality** (AR) pattern (figure 5.13) helps the user identify location and nearby features by visualizing them through superimposing digital information on whatever the user is looking at through their phone camera.

WHY

Studying a map takes time, and it's often difficult to orient yourself. It's hard to identify and locate points of reference in relation to your current location. This is especially true during time-sensitive tasks or in emergency situations in which objects are obstructed because of smoke, flood, snow, or nighttime darkness. In these situations, the sensors of the phone can detect more than just the current location—they also sense the direction the camera is facing and the inclination of the device in the user's hand. Combining these variables with reliable GIS data reduces the user's cognitive load and speeds up the decision-making process while the user is on-site at a location.

WHEN

AR is useful whenever features are directly related to the user's current location. It's easy to look through the lens of a camera and see where the nearest exit is, even if the exit is temporarily hard to find. Utility workers, for instance, can visualize obstructed assets and critical infrastructure, such as water pipes and electrical cables that are behind walls or buried underneath the street's surface. AR helps guide on-site workers in locating these otherwise invisible underground utilities.

Using AR in conjunction with other sensors such as the camera allows law enforcement to look at a suspect's car through the viewport of the phone, scan the license plate, retrieve the driver information, and display it right next to the car. People interested in real estate can look at a building and see the apartments available for rent, including their location, price, and contact information.

AR is best used outside so that highlighted features and points of reference are visible. It's more difficult to achieve a similar experience inside a building where walls block the view, and the device location may be less reliable.

HOW

Provide location-based labels, sometimes called billboards, with basic information about the feature or features in question. Keep the titles short and descriptive without superfluous wording or descriptions. Relate the location of the object to the user's location by displaying the distance in feet or meters. Avoid overlapping labels through prioritization and use callouts. Callouts that are further away from the current location may be displayed smaller to indicate their distance from the viewer.

Allow users to interact with an object by selecting it to open an info pop-up or info panel. The pop-up can show additional information such as photos, contact information, or reviews. Sometimes it is beneficial for users to switch the view from AR to a map to get the full picture or to show a small portion of the map side by side with the AR view. Common actions available in the pop-up are getting route directions, adding a note, using the object as a starting point for a workflow, or sharing the object with friends or coworkers.

Consider inevitable fluctuations in GPS location and data accuracy, which may result in imprecise representations of the object's location. The same applies to measurements calculated through AR that are mostly only approximations.

It is important to note that walking or driving while looking through the lens of the upheld phone is not only tiring but also dangerous. Therefore, the app should take precautions to remind or warn the user to watch their surroundings to prevent and avoid injuries.

EXAMPLE

AuGeo is an Esri Labs initiative to use ArcGIS data in an AR environment. While using the app, users hold their phone to see data from an ArcGIS point feature layer superimposed on their screen. Figure 5.14 shows how the app can help you identify where the ski lifts are located. It's easier to look through the lens of your readily available phone than studying a paper map and figuring out where the lifts are in relation to your current location.

Figure 5.14. Using Esri's AuGeo AR app to find the ski lift.

References

Android, *Documentation for App Developers,* https://developer.android.com/docs.

Apple, *Human Interface Guidelines,* https://developer.apple.com/design/human-interface-guidelines.

ArcGIS AppStudio, www.esri.com/en-us/arcgis/products/arcgis-appstudio/overview.

Chivite, I. 2017. "AuGeo: An Augmented Reality Mobile App for Your GIS Data." *ArcGIS Blog,* July 6, 2017. www.esri.com/arcgis-blog/products/3d-gis/3d-gis/ar-for-your-gis.

Hoober, S., and E. Berkman. 2012. *Designing Mobile Interfaces: Patterns for Interaction Design.* O'Reilly.

Chapter 6

Building single-purpose apps

Single-purpose apps focus on solving one problem well. They specify the job to be done and provide only the essential functions. This characteristic makes them easy to use and leads to a more focused UX. Single-purpose apps are often based on existing templates, configuration procedures, or builder workflows, which makes them fast to develop and deploy. This chapter introduces the most common single-purpose app types and walks you through important design considerations before using them.

App templates

The fastest way to build focused apps is to reuse existing app templates that provide the UI and functionality to solve a specific problem. In most cases, it's up to you to find, download, configure, and host the template. An alternative to this approach is to use a hosted solution such as ArcGIS Instant Apps.

Instant Apps provides a collection of focused app templates and an experience to configure them. Learn more at www.esri.com/en-us/arcgis.products/arcgis-instant-apps.

You can rely on the templates to be accessible and performant on any device. Instant Apps supports integration with your web maps and will host the app for you. Following are some of the templates available in Instant Apps:

- Basic and Minimalist allow you to create **full map** apps.
- Sidebar is a great starting point for a **partial map** app.

- Chart Viewer allows you to create a basic dashboard app.
- Slider is a quick way to show temporal data using a **timeline slider**.
- Nearby and Zone Lookup are examples for locator apps.
- 3D Viewer creates an app with an embedded **3D** map.

Locator apps

A locator app is sometimes referred to as a finder app and allows users to locate a site such as a school, pharmacy, or bank. Commercial websites often use a **store locator** to help their customers find the closest store location. Locator apps can be simple, single-page apps that are embedded in another website or a dedicated app that follows a workflow as outlined in chapter 5, "Designing for Mobile Devices." To build a simple locator app, you can use Instant Apps, which offers an easy-to-configure app template called Nearby. Nearby helps people find sites of interest close to an address.

Locator apps require an address as the main input but usually also allow users to share and use their current location. The app then shows the results in a list and side by side on a map with the ability to filter the results by additional criteria. Successfully using a locator app involves the following three steps:

1. **Find the locator.** Provide a global link to the locator labeled Find a Store, Store Locator, or Locations. For native mobile apps, promote the app for download from an app store.
2. **Use the locator.** Let users search by address, place, and current location. For each result, show distance from user's location, name, phone number, business hours, and actions to get directions or call.
3. **Get directions.** Create a route that uses the user's address as the starting point and the selected store location as the destination.

Navigation apps

A navigation app is a web-based or native mobile app that helps users find routes and provides directions. Navigation apps require users to define at least a start and an end point to calculate the route. They then use these stops and the selected travel mode to calculate the route and display live or turn-by-turn directions. Navigation apps are built for mobile users who are on the go and rely on **route** directions to reach their destination. They work especially well on native mobile apps with access to device sensors such as GPS or an accelerometer.

Various methods to select the start and end points exist. Although the starting point is most likely the user's current location, the following other positioning methods can be considered:

- Locate me: Access the user's current location to define the starting point.
- Location finder: Find a place by typing an address.
- Search: Search for offices, persons, assets, or any other object that has a location.
- History: Select from a list of previously visited or saved locations.
- Points of interest: Let users select from a list of well-known landmarks, such as cafeterias, libraries, gyms, post offices, or restaurants.
- Nearest facility: Provide a list of facilities such as restrooms, ATMs, kitchens, or first aid stations, and route users to the closest one.
- Select on map: Allow users to select a location by clicking in the map. Adjust this location by dragging the map underneath that point instead of moving the actual marker.

Provide a way for other apps to integrate with your navigation app so that they can prepopulate the starting point with an address or place.

Storytelling apps

A storytelling app, often referred to as a story, is a web-based app that allows you to share your maps in the context of narrative text and other immersive multimedia content. Stories are a helpful way to guide users through a potentially complex subject. Users can explore the topic by scrolling through the story, which advances the map to different points of interest (scroll to pan). Story-driven maps involving sophisticated data take care of map navigation and make it easy to immerse yourself in the story instead of dealing with the technical intricacies of map interactions.

Before you create a story—for instance, by using ArcGIS StoryMaps, which requires an ArcGIS organizational account, or StoryMaps, the latest storytelling tool, which uses a separate, non-GIS account—you need to know who your target audience is. This knowledge drives the tone of voice, vocabulary, and permissible level of detail. Next, you want to define the key takeaway of the story and draft the storyboard outline. Understand which content is already available to you and what is still missing, and aim to keep the content consistent in color and style. It's advisable to create and use a theme that follows your brand guidelines.

ArcGIS StoryMaps helps you tell remarkable stories with custom maps that inform and inspire. Learn more at www.esri.com/en-us/arcgis /products/arcgis-storymaps.

StoryMaps lets you share your most important ideas and experiences as interactive content featuring maps, media, and more. Learn more at www.storymaps.com.

When developing the story, follow these general guiding principles:

- **Start with a bang:** Use a powerful image and a meaningful title as the hook to your story.
- **Add a hero:** People love seeing people so consider using a person that your users can relate to, and then introduce tension to build the story.
- **Create a natural flow:** Introduce the big picture and then zoom into the details. Give the story a rhythm and repeat similar structures if possible. Keep it short and simple.
- **Add a call to action:** Now that you've told your story, what should your reader do next?

Dashboards

A dashboard is a web-based app that reports metrics about a subject or objective. Dashboards can be operational (what's happening now), strategic (track key performance indicators), analytical (identify trends), or any combination of the above. Users of dashboards include executives and people who need at-a-glance data reporting to make informed decisions. Interactions are typically kept to a minimum, and data visualizations such as maps mostly exist for reference purposes only.

A dashboard can be created by using software such as ArcGIS Dashboards.

Dashboards enables users to convey information by presenting location-based analytics using intuitive and interactive visualizations on a single screen. Learn more at www.esri.com/en-us/arcgis/products/arcgis-dashboards.

Guidelines for good dashboard design include the following:

- **Communicate effectively:** Keep the metrics simple, and focus the dashboard on its purpose.
- **Represent your data well:** Display only useful data, and choose the right representation.
- **Be consistent:** Follow clear and consistent value formatting, avoid unnecessary redundancy, and provide headings that are easy to scan. Keep the UI consistent in font size, margins, and visual appearance.
- **Keep it simple:** Each element on your dashboard must have a purpose. Avoid relying on interactions to convey meaning.

App builders

App builders are tools that allow nondevelopers to author compelling web apps by dragging premade functionality, so-called widgets, onto a canvas without writing a single line of code. The author will then connect these widgets with their own data and to each other and ultimately click the Publish button to generate a hosted web app that looks great and runs seamlessly on desktop and mobile devices. App builders such as ArcGIS Experience Builder also allow authors to create app templates that can be shared across their organization as an easy and fast starting point to share customized experiences.

Experience Builder empowers you to quickly transform your data into web apps without writing a single line of code. Learn more at www.esri.com/en-us/arcgis/products/arcgis-experience-builder.

There are two ways to create a new no-code app using Experience Builder: you can start from scratch on a blank canvas, or use a ready-made template that meets your needs. The former is a great opportunity to design a solution that perfectly solves your problem (refer to chapter 1, "Getting Started with Design") and uses the most suitable layout (refer to chapter 2, "Selecting the Right Layout"). This approach may require a little more effort to build the app experience but provides the greatest flexibility. The latter approach works great if you know of a template that closely resembles your desired app design. You need to only open the template, connect your data, and save it as a new solution. Selecting the right template starts with knowledge about your target audience (refer to personas in figure 2.1 in chapter 2). Based on the persona, the following Experience Builder templates are recommended:

- **Analyst:** Choose from the ArcGIS Web AppBuilder WAB Classic section, such as Foldable, Launchpad, or Jewelry Box.
- **Executive:** Dashboard templates such as Monitor work best.
- **Public:** A wide variety of templates options exist. Pick from Map Centric and WAB Classic sections. For web pages, use Introduction or Seeker; for websites, use Scenic.

Patterns

The patterns in this chapter are equally relevant and important for single-purpose apps and other types of apps. The first set of patterns are workflow driven and involve location to improve the user experience. This set includes the following patterns:

1. **Location list**
2. **Store locator**
3. **List and details**
4. **Search and edit**

The next two patterns are important concepts used by storytelling apps:

5. **Scroll to pan**
6. **Extent-driven content**

The following patterns are described online and include two routing patterns that describe workflows for using location to position the user and provide directions:

7. Route positioning (www.mapuipatterns.com/route-positioning)
8. Route directions (www.mapuipatterns.com/route-directions)

Some single-purpose apps specialize on narrow tasks. They strive to solve one problem only, which helps optimize the task and market its solution:

9. Meet in the middle (www.mapuipatterns.com/meet-in-the-middle)

Location list

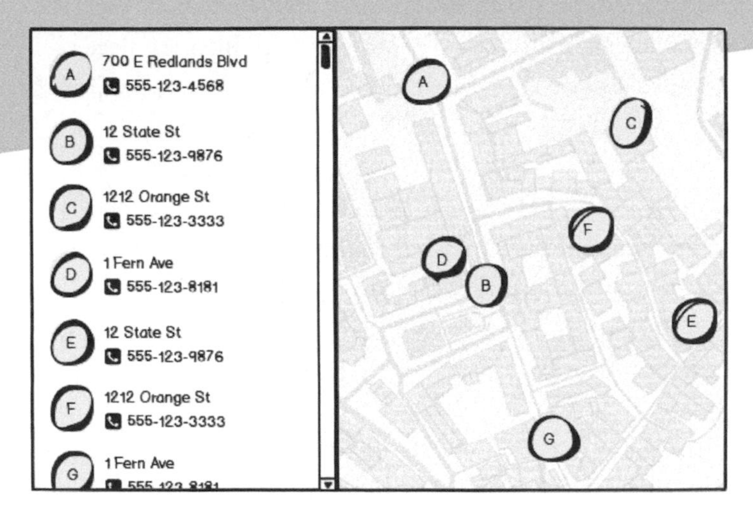

Figure 6.1. Location list pattern.

WHAT

The location list pattern (figure 6.1) consists of a list of locations, often referred to as sites, and is visualized by markers on a map. A key, like a legend, establishes a relationship between locations on the list and in the map.

WHY

Both elements, the list and the map, have their strengths and limitations, respectively. Map markers show the location of an object in relation to other geographic features. Providing spatial context is their main strength but they are limited to displaying only one or two attributes through color and size. One way to solve this limitation is to add labels, but this contributes to undesired clutter in the map. Lists, on the other hand, are great for displaying and manipulating tabular data but lack the spatial context. Displaying the list side by side with the map and providing a key that relates the two eliminates many of these shortcomings. Users can use either component to explore the sites and interact with the locations.

WHEN

Location list is a pattern commonly seen in consumer apps that allow users to search for places such as hotels, restaurants, or stores. It's an integral piece of any locator app. The **store locator** is a specific and widely used variant of the locator app.

Analysts, managers, and operators are used to working with spreadsheets. Location list helps them sort, filter, and parse through rows of tabular data while simultaneously identifying objects or clusters of objects on the map.

HOW

Location list works best for apps with few sites and where the distance from the user's location is of high importance, such as a **store locator**. Provide a list and a map in which the list component can be either superposed to the map for **full map** layouts or juxtaposed for **partial map** layouts. Establish a relationship between the list item and the **marker** by using a unique key, such as a number or alphanumeric character, for each item. Use data brushing that lets users interact with a data item in one view while the corresponding item in another view lights up simultaneously. In the case of the location list, hovering over a data row in the list will highlight the marker, and vice versa. If the corresponding item isn't visible in the current view, nothing will be highlighted.

The list component often follows the appearance of a multicolumn, tabular grid, but simpler, single-column layouts are possible and often visually more pleasing. Regardless of the design, content should include one or more of the following elements:

- **Attribute values:** Names, links, addresses
- **Status indicators that are often related to workflows:** Pending, assigned, open, closed, rejected
- **Metrics or indicators that support the user in planning:** Acres burned, wildfire percentage contained, miles to the user's current location
- **Actions:** Zoom to the item, view item details, edit attributes

The following common interaction patterns are related to the location list and can be found throughout the book:

- Extent-driven content: Answers the question whether changing the map extent should modify the results shown in the list
- List and details: Explains the intricacies of interacting with a list item, viewing and manipulating complex data
- Store locator: Describes a use case that is widely used in retail applications
- Spatial filter: Illustrates how filtering by geography will affect the map and the list
- Feature selection: Creates a collection of features to be displayed

EXAMPLE

The Geospatial Insurance Consortium (GIC) Virtual Inspection Studio app (figure 6.2) shows insurance claims that require review by a virtual adjuster. The claims are shown as a list in a panel on the left side and by using round markers on the map on the right. Data brushing establishes a visual relationship between the two so that hovering over a claim will add a gray background to the list item and a blue circle around the marker on the map. Each list item indicates the claim status, the owner name, claim ID, property value, and actions to see details or delete the claim. An example of the claim details view can be seen in the focus on the subject pattern. The app makes smart use of extent-driven content, in which a check box labeled Visible On Map allows users to opt out of the default behavior to show only claims that are currently visible in the map.

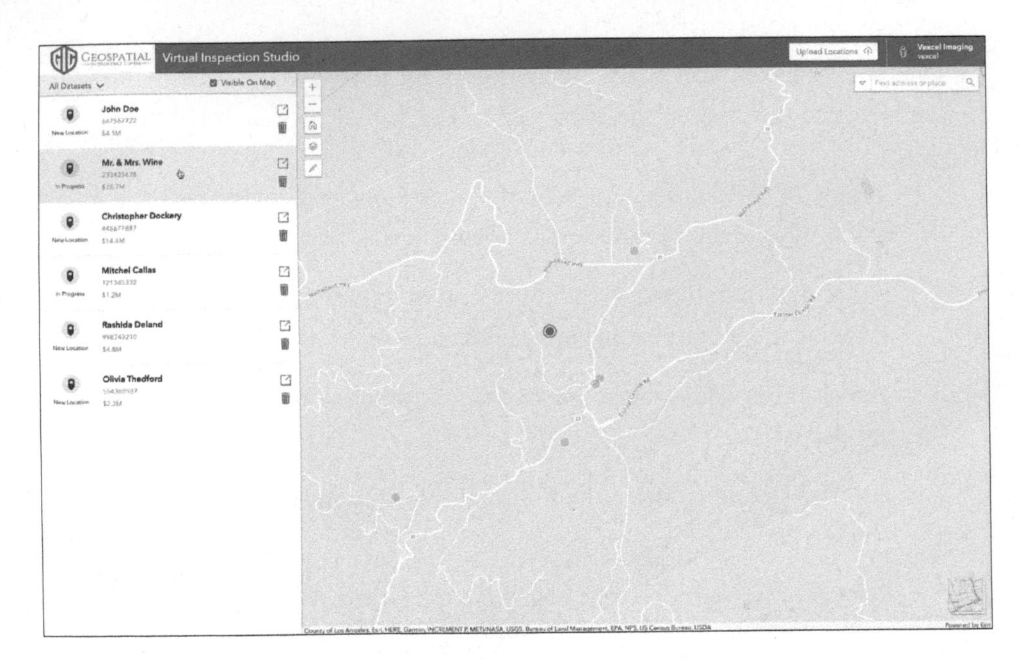

Figure 6.2. GIC's Virtual Inspection Studio relates the list items to the locations on the map.

Store locator

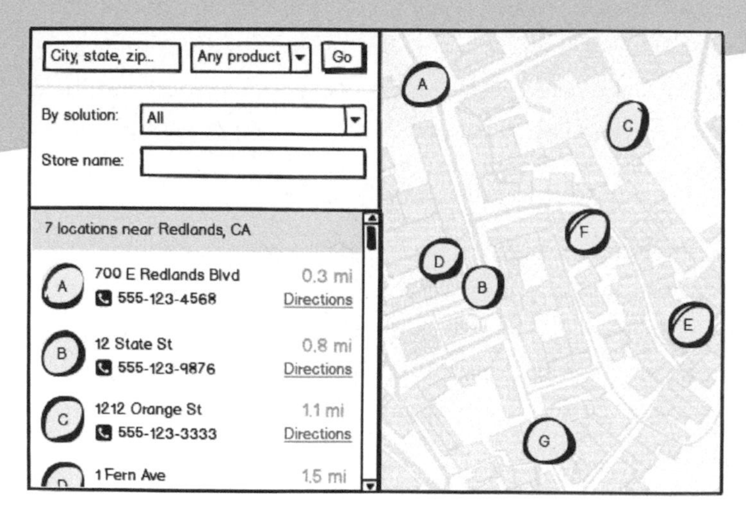

Figure 6.3. Store locator pattern.

WHAT

The **store locator** pattern (figure 6.3) finds physical store locations closest to a place, address, or the user's current location. It is important for companies to provide a store locator that helps their customers easily find the most suitable real-world store location.

WHY

Most businesses operate on the internet, but their customers still live in the physical space and may need to visit brick-and-mortar locations such as a store, bank, office, dealer, or facility. If they cannot locate a location that is close to them, the business loses money. The company's website must provide an easy way to find and open the store locator itself, and the store locator workflow should be optimized to find the most convenient store that offers the services needed.

WHEN

Any business with physical store locations or services should provide a store locator. The number of stores will dictate how to present the location information. If the business has only one location, it's best to display the address, phone number, business hours, and a link to get driving directions directly on the home page or Contact Us page. For few locations, provide a static location directory that can be accessed from every page throughout the app or site. Once your business has 10 or more locations, resort to implementing a dynamic store locator.

E-commerce sites with physical locations, such as big brand or shipping stores, may choose to embed a simplified version of the store locator directly into their workflows. Based on the user's current location, the closest store is preselected. Product listings and business hours are then based on availability for that location. Users can update the default location by entering more exact location information, such as a zip code.

HOW

A store locator can be successful only if users can find it first. Provide clearly labeled links such as Find a Store, Store Locator, or Locations, and avoid branded terms. Add a link to the locator from all the pages within the site, preferably in the header or footer sections. Consider placing additional links on strategic pages such as the shopping cart page to find pickup locations or product pages to check in-stock inventory.

The process to locate a store should provide the following steps:

1. Enter location information. Use **location finder** and **locate me** to identify the desired location.
2. Optionally, apply **attribute filter** to limit the locations to meet certain criteria such as specific services, products, or availability.
3. Activate **search** by clicking a button labeled Search, Find, or simply Go.
4. Add a **marker** and center the map around the location. Use a street map as the **basemap** layer.
5. Show results in a list and on a map. Use **location list** to show locations by distance, and include important attribute information, such as store name, phone number, and business hours.

6. Refer to location list earlier in the chapter for additional best practices on how the list and map should behave.
7. Retain the selected search and filter options so it's easy to refine and search again.

The two main edge cases to consider are that the locator returns either zero results or too many results. The following suggestions can help you optimize the user experience:

- **No results found:** Consider increasing the search radius to check whether more results become available. This approach works well for specialty stores or fine-dining choices in which users are willing to drive longer distances. It's less likely to be a good choice for commodities such as getting a cup of coffee. Alternatively, inform the user that no results were found and offer a way to shop for the same product or service online.
- **Too many results:** Start with a default search radius that works well for most locations and without overwhelming the user. Use implementation techniques such as lazy loading to save bandwidth by showing only a few locations at a time. Reduce the results by offering attribute filters, such as services offered, product availability, or business hours.

EXAMPLE

The mock-up in figure 6.4 shows an app to find libraries. It illustrates how the app uses a simplified store locator embedded in the page header. On start-up, the app remembers any previously selected site or locates the closest one. The trailing down arrow indicates that this option can be changed, and as the user clicks on it, a panel to select another library location slides out from the left. The top section of this panel includes a locate me link and a location finder, followed by a list of nearby sites. The currently selected site is marked as such, but the user can easily change to another site by clicking the Set as My Library button.

The app also offers a more comprehensive store locator that can be opened through a link in the top-right corner labeled Find Locations. This store locator opens a dedicated page with a list of libraries side by side with a map.

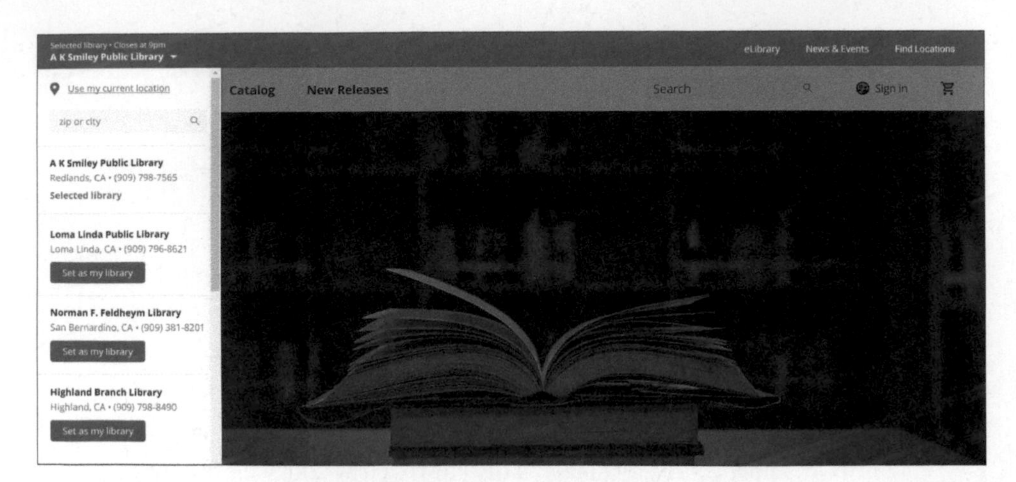

Figure 6.4. App to find libraries uses a simplified store locator to select the library of choice.

List and details

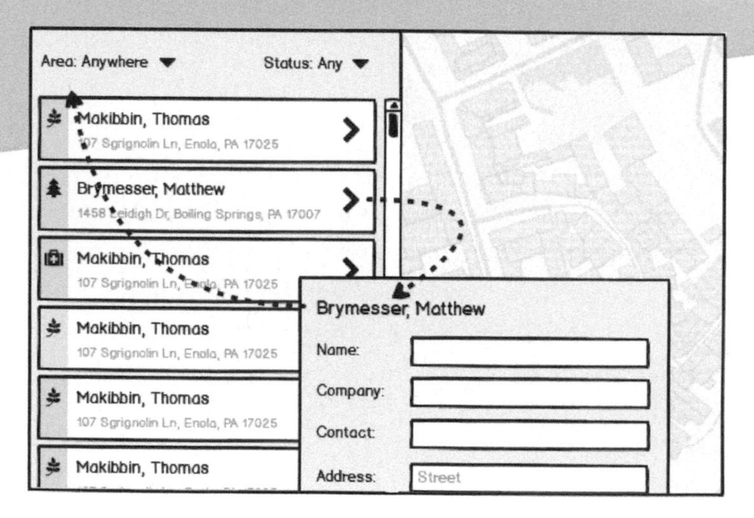

Figure 6.5. List and details pattern.

WHAT

The list and details pattern (figure 6.5) is related to the location list pattern. Both patterns share the same layout and use similar interaction patterns such as data brushing. The main difference is that with list and details, a user can select an item in the list to view and edit its attributes, or details.

WHY

A list is a series of objects that reflects data retrieved from a database or service. Most lists are displayed in the form of a simple list that displays the information in a card or a more complex table that consists of rows with multiple columns. Most list items show only a few attributes of the whole dataset, such as title, name, and other key fields. The underlying data is usually more complex, and there isn't enough space available to show more than a handful of attributes. It's almost impossible to edit this data inline or expand a row into a long form of attribute values. If users are interested in seeing or editing the details of a list item, they can navigate from the list to the details view. The details view helps the user focus on their tasks, such as performing an analysis or updating attribute values. Once the task is completed, users navigate back to the list. This pattern works well for items that have long or complicated attribute forms. It helps focus the user on the selected item. It also affords embedding map interactions, such as editing the geometry into the workflow.

WHEN

List and details is used in apps for situational awareness, operations, or asset management. The two types of apps that benefit most from this pattern are single-purpose and enterprise apps. Single-purpose apps can focus users on managing assets, projects, incidents, or any type of features that require a lot of attributes to be edited. Enterprise apps may use this pattern to show lists of clients, assets, or activities with their status, address, contact information, time stamp, or location.

HOW

Provide a list of items side by side with the map. The list is often sortable, filterable, searchable, driven by the user's privileges, and populated through services from back-end servers. Keep the list simple by displaying only the most important columns and removing unnecessary design elements such as separators, row striping, shadows, or even cell borders. Offer a button to create a new item and place it toward the top of the list.

It's not always necessary to provide a details view. For instance, simple attribute updates can be accomplished using inline editing or by opening an info pop-up. A dedicated details view is required for more complex data display or longer forms. The list view and details view are usually contained within the same panel. Users can navigate to the details view by clicking an item in the list and navigate back to the list view by clicking a simple back button on the details view. If the details view holds a form, you should provide buttons to save and cancel instead of the back button. You may also consider adding Preview and Next buttons or arrows to navigate between records in a linear fashion. This will reduce the need to toggle back and forth between the list and the details.

After an item is selected from the list, the app should remember the current extent and zoom the map to the item's extent showing the necessary level of detail for the user to analyze the item and interact with it. Upon the user returning to the list, the previous extent should be restored so the context isn't lost.

It isn't uncommon for the details view to require more horizontal space than the list view. It's okay to adjust the width for each view if this mechanism is applied consistently. Consider adding a button to collapse the panel. Place the button at the vertical center so that it overlaps the map. The button collapses the panel regardless of the current view.

EXAMPLE

The BlackSky Tasking for ArcGIS Online app (figure 6.6) provides a convenient way for users to get up-to-date satellite imagery delivered in as little as 90 minutes. After the area of interest is specified, the app tasks a satellite to capture new imagery at the earliest available possibility. After successful collection, the imagery is ready to be transferred to ArcGIS Online. The order history section gives users the opportunity to review details of past and current orders. When a record is selected from the list on the left side, the app follows the list and details pattern to show more details while simultaneously zooming the map to the extent of the selected record. Depending on the delivery status, users may be offered additional actions, such as retasking the satellite or accepting the order to be moved to their ArcGIS Online organization. Closing the details view returns to the list and zooms the map back to the previous extent.

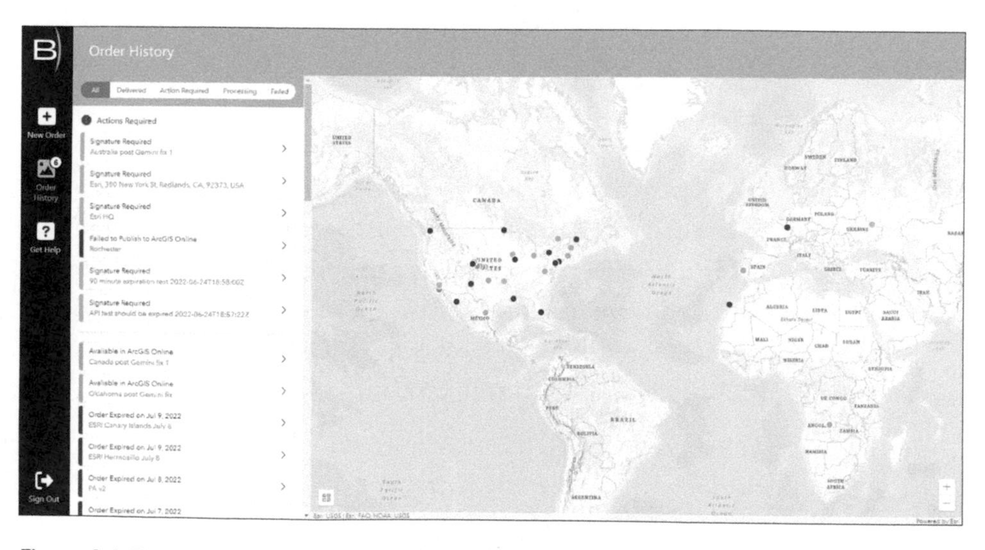

Figure 6.6. The BlackSky Tasking app allows users to order satellite images of a specific region and have them delivered to an ArcGIS Online organization.

Search and edit

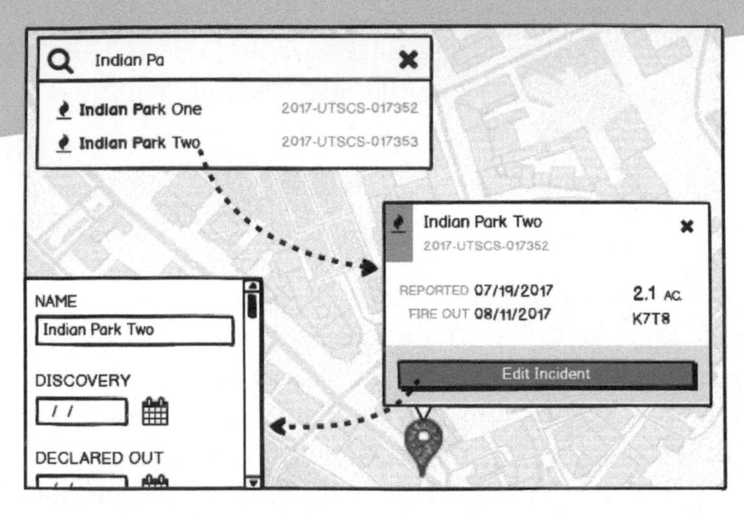

Figure 6.7. Search and edit pattern.

WHAT

The search and edit pattern (figure 6.7) helps users find features that need to be updated or modified quickly. The typical use of search is to find an item and display its location and attribute values but users can also perform a search for an item so they can edit or even delete it.

WHY

Discoverability is the number one need for many single-purpose apps. Browsing through lists of data may work well for small amounts of data, but as datasets grow larger in volume, finding specific content quickly and reliably becomes critical for the UX. Apps that manage large amounts of data therefore require a search to find items quickly. Viewing the selected search result item is the dominant driver for a search in most cases, but providing a button to edit the item helps streamline the editing process.

WHEN

Single-purpose apps use search and edit to find features, objects, or records so they can be updated quickly. Users can be operators receiving phone calls from customers, subscribers requesting updates to their accounts, or field crews updating assignments or work orders. Using search and edit, which is a good example of a task-oriented pattern, is faster than selecting the correct layer, zooming to the area of interest, and visually identifying the feature on the map. It's especially useful for apps that manage large or unfamiliar datasets.

HOW

Implement a federated search that displays type-ahead suggestions. Once the user selects the desired search result, the map zooms to that feature and shows the item content in an info pop-up. The content should include the name or ID, some key attributes, potentially some statistics, and an action button that reads View Details, Open for Edit, Update Record, or similar. Upon clicking this action button, the full item details open in a separate panel. Full map may choose an info panel, whereas partial map can use the content area of the side panel.

The search and edit pattern has similarities to the list and details pattern. Consider the following deviations between the two patterns:

- After selecting a search result, it can be useful to keep the results list open so that users can jump between results until the desired item is found. Alternatively, provide an easy way to recover the search by keeping the search term persistent.
- Structure the content of the info pop-up so that users can assess quickly whether the selected feature is indeed the correct one.
- If the app uses location list, consider showing the same item card in the list and in the pop-up.
- Avoid opening the item details in edit mode right away. Offer an easy way to trigger edit mode instead. Provide explicit save and cancel actions that both navigate back to the item details in view mode.
- Besides editing the item, consider alternative actions such as close, delete, transfer, attach, share, or analyze to manipulate the item.

EXAMPLE

Figure 6.8 shows a mock-up of an app for managing farmland ownership accounts. The app uses a **partial map** layout to show the landownership accounts in a list side by side with the accounts in a map. Users can filter the list by area (**spatial filter**) or account status (**attribute filter**) and toggle to the details of an account following the list and details pattern. The easiest way to find an account is to **search** by name, address, project number, or farm number. Selecting a search result zooms to the farm and opens an **info pop-up** with basic information about the account. At this point, the app senses that the user's intent most likely is to modify the account, and that's why the pop-up shows three action buttons: Edit Details, Transfer, and Close. Edit Details has the same effect as if the user had clicked on an item in the list and opens the details panel in edit mode. Transfer starts a **task-oriented** workflow to transfer the account to another person. Close is a drop-down list that allows users to choose from different ways to close the account, which changes the status from Active to Closed.

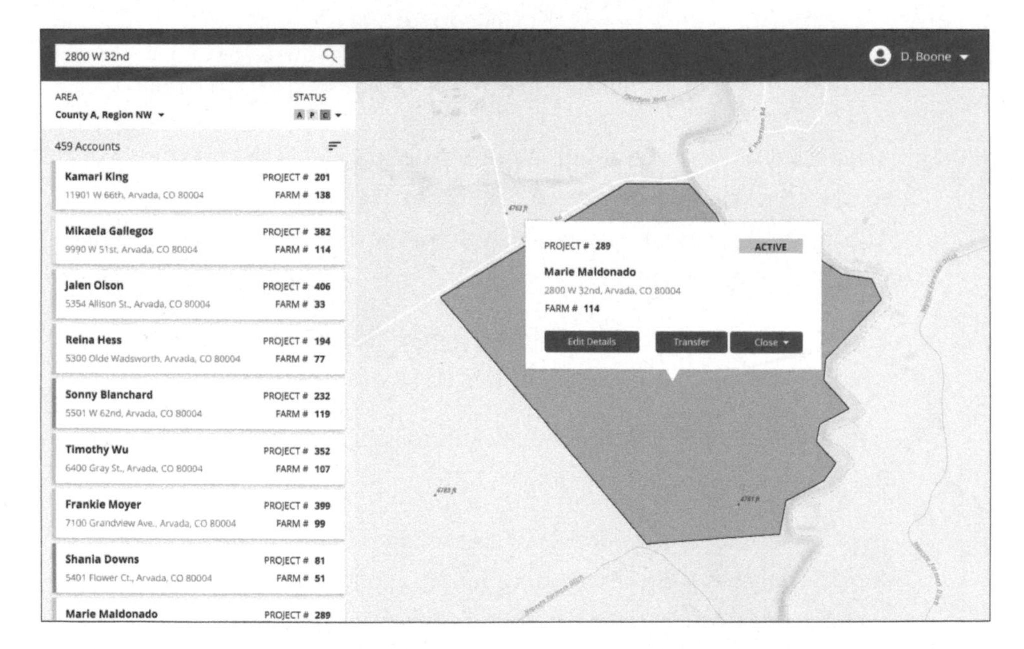

Figure 6.8. Search and edit allows users to search for an account and start a task-oriented workflow.

Scroll to pan

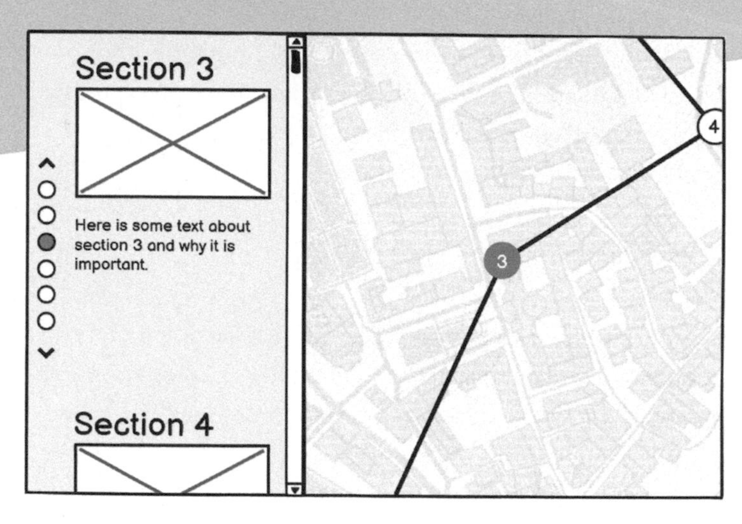

Figure 6.9. Scroll to pan pattern.

WHAT

The scroll to pan pattern (figure 6.9) is a technique to walk users through a series of maps one step at a time. It uses the browser's scroll bar to move through a series of predefined map content.

WHY

The existence of a scroll bar indicates that more content is available below the fold, as in a newspaper fold, which separates the first screenful of material from the remaining content. Users have learned that scrolling up and down through a page reveals the hidden part of the page. Using the mouse or keyboard has become second nature to them. Builders can leverage the user's familiarity with the scrolling technique to create smooth transitions between a series of predefined maps, images, or other media.

WHEN

Scroll to pan is widely used in storytelling apps to support the narrative by walking the user through the data in the way the author intended to tell the story. This removes the burden of the user to explore the map and figure out for themselves what is important. Instead, the app takes the user from the big picture of the data to specific insights or connects a series of events or places that are related to each other. For instance, the app might want to introduce users to the series of World Cup soccer stadiums or tell a story about the ruins along the Inca trail, a multiday hiking tour in the Peruvian Andes. As the user scrolls the page, the story consecutively shows information about each stadium or ruin, respectively, side by side with its location.

HOW

Users understand how to navigate an app following familiar scrolling patterns. Use this knowledge to update the map choreography, such as pan or zoom, as new content becomes available. Show narrative panels, sometimes also called sidecar, either superimposed on the map or on a panel next to the map. Their purpose is to narrate the story and explain what is happening. They also draw attention to the map visualization and establish a relationship between map layers.

The following approaches can be used to update the map choreography:

- Pan and zoom the map extent to guide users through a series of places.
- Toggle layers on and off to smoothly animate from one view to the next as a means of comparison.
- Progressively reveal layers to build up the complete picture, which is less overwhelming than showing all the layers at once.

Anticipate the situation in which users don't immediately know that they are supposed to scroll the page. In these situations, consider adding the following visual aids to indicate that there is more content to explore below the fold:

- Intentionally cut off content elements such as images or a paragraph of text.
- Place a down arrow toward the bottom of the screen.
- Add a carousel navigation such as clickable dots that serve as page jumps and arrows to move back and forth. This navigation should be sticky along the side or at the bottom of the screen.

- Animate a small pull-up motion.
- Autoscroll the story at constant speed. This approach works well on static displays or when embedded in an operational dashboard.

Scrolling usually happens in a continuous fashion, but sometimes the page may move in discrete increments—for instance, by clicking on a dot on the carousel navigation.

EXAMPLE

Mapping the COVID-19 Pandemic (figure 6.10) is a story made using ArcGIS StoryMaps software about the emergence and spread of COVID-19. The app uses scroll to pan to illustrate the path and impact since its outbreak. It starts with a map of its origins in Wuhan, China. As the user continues to scroll down the page, the map zooms out and pans to different parts of the world. This automated map behavior goes hand in hand with the development of the story that shows how the outbreak caused actions to quarantine the region, suspend domestic and international travel, and deal with first reports of cases and fatalities coming in. It also shows how the virus started to spread into other countries in Europe and eventually the United States. Descriptions of these events pass by on the left side while the map pans and zooms to new regions and progressively reveals analytical layers that support the storyline.

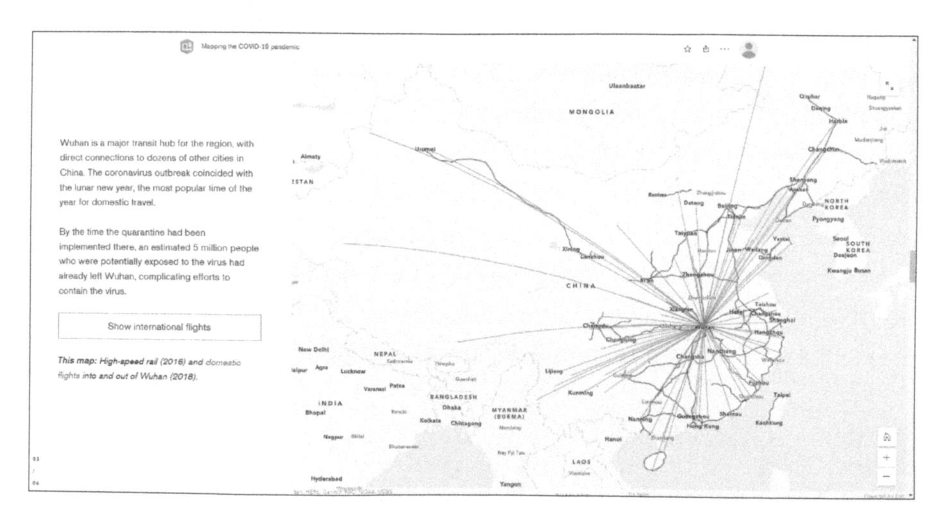

Figure 6.10. The story shows how the novel coronavirus emerged and started to spread across the globe.

Extent-driven content

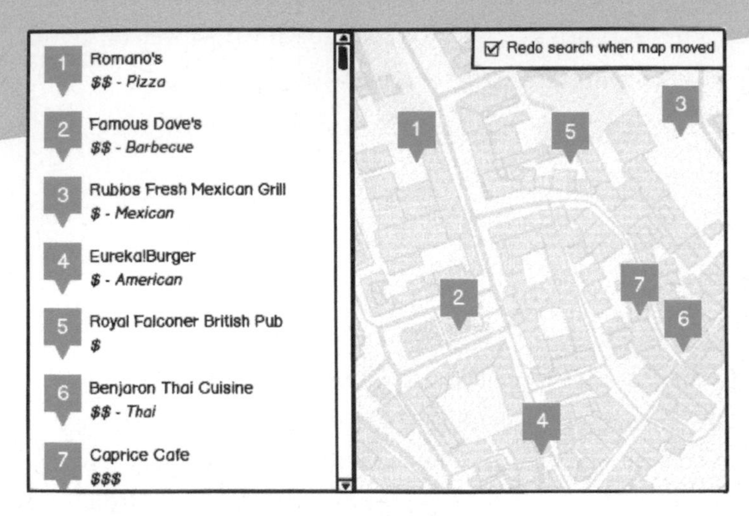

Figure 6.11. Extent-driven content pattern.

WHAT

The extent-driven content pattern (figure 6.11) synchronizes content in the UI with the map extent. It describes how the map can be used as a dynamic spatial filter that updates other content on the page as the map extent changes.

WHY

Data can be visualized in many ways. The most common visualization components are charts, tables, maps, and simple indicators. Each component provides a unique viewpoint into the data and allows users to adjust the display to their needs—for instance, lists can be sorted, and maps can be zoomed and panned. Manipulating each component changes the center of interest without affecting other components deployed on the page. This behavior is expected because users do not want to see lists and charts update constantly after each miniscule pan event. Extent-driven content describes an exception to this rule.

WHEN

The most important application of extent-driven content is exploration of places, sometimes also referred to as short list. Users want to explore points of interest, such as hotels, restaurants, or landmarks, within a given area. While in this mindset, they are interested in seeing only data that lies within the current map extent. Navigating the map controls the content displayed in related visualization components on the page such as the list. The list therefore provides visual feedback by reflecting the places currently visible on the map.

HOW

Provide a partial map layout with markers that relate to points of interest. Add other visualization components such as a list, chart, or indicators adjacent to the map. Bind the data to the same source and synchronize the components. Use the current map extent as a spatial filter that dynamically updates as the user zooms or pans the map.

Navigating the map without constraints comprises the following challenges that you need to consider:

- **User zooms in too far:** Zero results are available. Provide empty states for related visualization components.
- **User zooms out too far:** A large amount of irrelevant data is displayed and downloaded. This navigation behavior is often an interim state to find and narrow in on another geographic location. There is little value in showing all available points at smaller scales. Consider restricting the data download to a predefined buffer around the center of the map. Alternatively, use cluster marker to hide individual markers.
- **User pans away from their current location:** This behavior is a common side effect of any exploration behavior. Add locate me to re-center on the user's location.
- **User pans outside the area of interest:** If your data is restricted to a certain area such as a city, the user may unconsciously zoom beyond these bounds. Consider notifying the user of that circumstance, and provide a home button to reset the extent to its original bounds.

As stated earlier, synchronizing the map extent with other content is often not the expected interaction behavior. Consider providing a method to opt in or opt out of this behavior by providing a check box labeled Redo Search When Map Moved or similar.

EXAMPLE

The Australian Tree Crop Map Dashboard is an initiative led by the University of New England research team from the Applied Agricultural Remote Sensing Centre (AARSC). The app (figure 6.12) summarizes the commercial horticulture tree crops in Australia, such as avocados; citrus, macadamia, and mango orchards; banana plantations; and olive groves. Users can review summary statistics for each tree crop in an **info pop-up**. Depending on the map scale, the statistics are calculated by the selected state at small scales, local government areas at medium scales, or tree crop polygons at large scales. Additionally, the app follows the **extent-driven content** pattern to update the crop indicators located underneath the map. The indicators show total numbers for each crop within the current map extent. These numbers update as the user zooms or pans the map and are independent of the data displayed in the pop-up.

Figure 6.12. Australian Tree Crop Map Dashboard ties the current map extent to the cumulative indicators underneath the map. Source: University of New England.

References

Cooper, Thomas. 2019. "Choreograph Your Maps with ArcGIS StoryMaps." *ArcGIS Blog*, November 6, 2019. www.esri.com/arcgis-blog/products /arcgis-storymaps/mapping/choreograph-your-maps-with-arcgis-storymaps.

Chapter 7
Common mistakes and how to avoid them

The purpose of this chapter is to improve your design literacy and equip you with the knowledge to foresee and prevent wrong actions that emanate from faulty judgment, inadequate knowledge, or not paying attention. The chapter describes common mistakes so that you recognize them before they become a problem. Consequently, you'll be able to combine these insights with suggested remedies to deliver better apps. After reading this chapter, nobody should say anymore, "I didn't know" or, even worse, "I know it's bad, but I don't know how to fix it."

Cost of design mistakes

How much does it cost to create a good UX? Is it worth the investment? These questions are difficult to quantify and often lead to the following answer: "We don't have time for that." As a result, solutions to newly arising problems will often be resolved by taking shortcuts, following opinions, or applying popular trends. These approaches lead to suboptimal solutions and bad UX, which leads to the question: What's the cost of bad UX? The impact of neglecting to invest in UX design can range from mere inconvenience to loss of productivity and a decreased conversion rate—that is, the percentage of users who take a desired action, such as completing a task, submitting a form, or making a purchase. Users spend more time completing their tasks, grow frustrated and angry, and may ultimately even abandon your app. Complicated and unfamiliar solutions result in faulty analyses, incorrect assessments, loss of money, and, in extreme cases, such as with public safety or emergency response, loss of life.

Frequent mistakes

The most frequent problems users encounter in maps fall into the following categories:

- **Inadequate visual salience:** Important information goes unnoticed.
- **Ambiguous affordances, or cues that hint how users can interact with an object:** Users notice UI elements but don't understand how to use them or have wrong expectations about their functionality.
- **Low information scent, or signals that users will find the content they want:** Users cannot find or understand what to do next.

Many of these problems can be solved by studying and applying the patterns outlined in this book. The following list is meant to highlight frequent mistakes and link them to the appropriate patterns for further reading:

- **Tendency to overvalue the map:** Maps are difficult to read and navigate. A simple **location list** is often easier to browse than finding the same information on a map. Use **full map** layouts sparingly. Often, **partial map**, **reference map**, or even **no map** are better options.
- **Starting with full extent:** People's area of interest is mostly where they currently are, live, or work. A **landing page** helps collect location information through **location finder** or **locate me** before displaying the map.
- **Too much data:** Receiving feedback that "the map feels crowded" is a sign that you fell prey to data overload. Avoid "drowning in a tsunami of data" by applying solution strategies described in chapter 4, "Dealing with Complex Data."
- **Forgetting about mobile devices:** Merely creating a compact version of your website is no longer acceptable. Chapter 5, "Designing for Mobile Devices," explains in depth the challenges and opportunities of mobile design.
- **Users aren't sure where to start:** Put yourself in the user's shoes. Would you know what to look at? Add a **focal point** to highlight the center of interest or activity.
- **Missing workflows:** This is a problem often found in widget-driven apps (**kitchen sink**) in which users must figure out the right sequence and combination for turning the right knob in the right place at the right time. Provide **task-oriented** workflows that align with the user's goals.

- **Poorly designed pop-ups:** It's easy to forget polishing **info pop-up** content. As a result, some pop-ups show poor default values, too many attributes, and irrelevant fields, such as ID values or coordinates. Labels should be cleaned up to avoid long, cryptic, and all-uppercase values with underscores as separators.
- **Strange-looking map:** Often the visual appearance of the map is different from the rest of the app and feels as if it were placed by accident. Create a custom **basemap** and adjust symbology to follow consistent visual style.
- **Gesture ambiguity:** **Embedded map** offers solutions to avoid unwanted panning behavior while users scroll through the page.
- **Splash screen on start-up:** Don't interrupt the user experience unless there is a good reason to do so, such as a legal requirement to acknowledge terms of use.

Patterns

Although you may not have heard of the names of the following patterns, you almost certainly have encountered their problems at least once in your professional career. It's usually easy to spot these problems and almost equally easy to avoid them by implementing the proposed solution.

1. **Kitchen sink**
2. **Desert fog**
3. **Data-ink ratio**

Other solutions that are available online include the following:

4. **Overview map** (www.mapuipatterns.com/overview-map)
5. **Home button** (www.mapuipatterns.com/home-button)
6. **Placemarks** (www.mapuipatterns.com/placemarks)

Kitchen sink

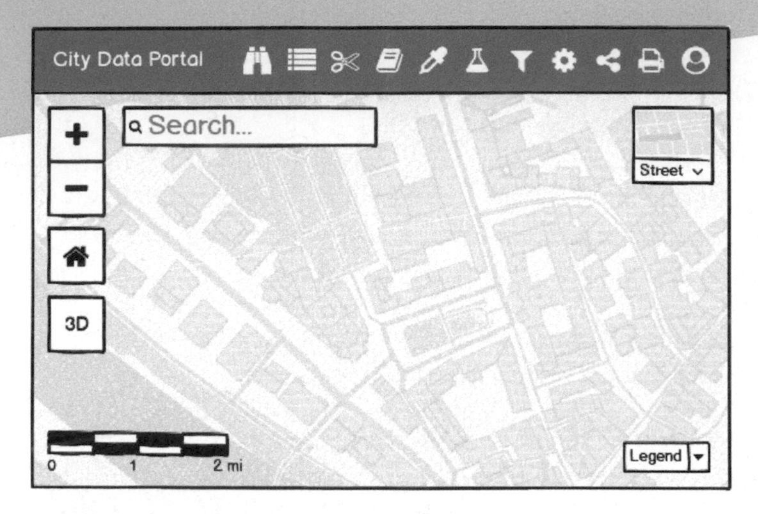

Figure 7.1. Kitchen sink pattern.

WHAT

The **kitchen sink** pattern (figure 7.1) showcases all conceivable functions on one page. The term *kitchen sink*, sometimes also referred to as Swiss army knife, is often used for apps that try to cram too much functionality and data into one interface. The result is a UI that appears busy, cluttered, less appealing, and not as user-friendly.

WHY

Mapping platforms offer a variety of useful features and ready-to-use widgets. Many of them are free and easy to include. It's therefore tempting to add yet another great feature to the app to be developed. The fallacy of that thinking is that each element added to the UI comes with a cost to its usability. As described in Hick's law, the time required to reach a decision will increase logarithmically with the number of choices available. The newly added feature that was meant to benefit the user may end up increasing information overload instead. Users will find it more difficult to find the right tool or follow a clear workflow. As a result, they will feel overwhelmed, frustrated, and, in the worst case, give up and abandon the app.

WHEN

Kitchen sink apps are best used to introduce users to a new platform, framework, library, or application programming interface (API). They are great for showcasing or demonstrating the entirety of its capabilities. For instance, developers who want to learn more about a product or platform can access and experiment with all the functionality available. Another way to use a kitchen sink app is for users who are familiar with the environment to walk their audience through a series of functions during a demo. Avoid kitchen sink for any other app because it is difficult to learn and can be confusing and overwhelming.

HOW

Provide a UI that allows users to access all the functionality through a series of buttons, tabs, or menus. Structure the functions by API reference, by category, or in another logical order.

Avoid kitchen sink unless your app falls into the category of demonstration or showcase. Rather than exposing a slew of functionality that is hard to understand and navigate, validate requirements against true end-user needs and focus the app on workflows accordingly. Limit functionality whenever possible and consider progressive disclosure of features and content.

Provide workflows such as **task oriented** or **search and edit** that support the user with their tasks.

EXAMPLE

The app in figure 7.2 uses kitchen sink to expose all the widgets available in ArcGIS Web AppBuilder. The purpose of this app is to showcase the breadth of functionality available in the builder tool, and that's why kitchen sink works well. Another assumption is that the user is either inherently familiar with the stack of tools or wants to explore its functionality randomly. It's not recommended to use kitchen sink for any other production-ready, user-facing app.

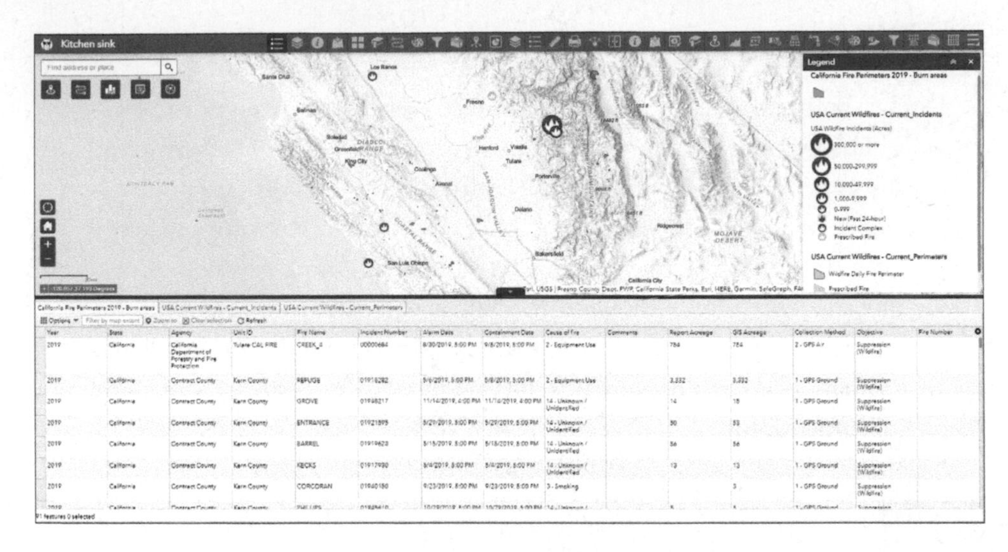

Figure 7.2. Kitchen sink app exposes all possible features for demonstration purposes.

Desert fog

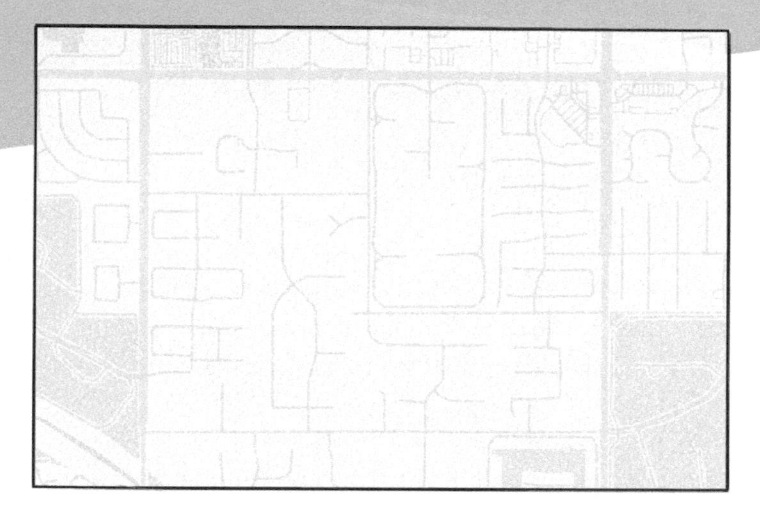

Figure 7.3. Desert fog pattern.

WHAT

The **desert fog** pattern (figure 7.3) offers no discernible UI patterns such as **search** or **home button** and no orientation clues on the map, leading to a loss of orientation for map users. Without these elements, it is increasingly easy to get lost in multiscale systems.

WHY

The reason users interact with maps is to explore data that is spread across vast extents and multiple scales. As users navigate unfamiliar geographic areas, they can easily lose their sense of place and start feeling trapped. Without additional information or clues that help users recover their sense of identity and get back on track, they will become "lost in the desert fog"—hence the name of the pattern. The same undesired outcome can occur after users choose an interaction by mistake or realize that their action wasn't what they wanted or needed. To avoid the desert fog problem, builders must consider the usability heuristic "user control and freedom," which states that great apps need to find the correct balance between enabling users and avoiding unwanted outcomes. The goal is to make people feel that they're in control and that there is an easy way to recover. This fosters a sense of freedom because users realize they don't need to be afraid of interacting with the UI.

WHEN

Desert fog can occur frequently when panning or zooming in an area in which there are no features that might help orient the user. Almost every map app must therefore provide one or multiple solutions to prevent the sense of losing orientation.

HOW

Various solutions can be used to avoid the desert fog problem. Some are related to the map and others to the app. The map plays a crucial role in communicating place, so you should consider the following cartography-related solution approaches:

- **Labels:** Make labels clear and legible with high contrast to the basemap layer. Consider adding a halo that resembles the dominant background color to ensure readability regardless of the underlying layers.
- **Key features:** Add visually salient or well-known features, such as landmarks, highways, rivers, or mountains. These features help users retain orientation during map reading.
- **Reference layers:** Include reference layers such as state or county lines.
- **Visual cues:** Consider adding other visual cues such as grid lines.

Most of the app-related solution approaches use components that are superimposed on the map. The following list describes the patterns that help avoid desert fog through the app:

- **Home button**: Restore initial map extent to a well-known area.
- **Locate me**: Re-center the map to the user's current location.
- **Placemarks**: Provide shortcuts to important places, such as US territories, that would require a lot of navigation otherwise.
- **Overview map**: Tell users where they are in relation to a wider view.
- **Location finder**: Search for a place as a final escape hatch.

Finally, avoid ambiguous affordances that would confuse users about the correct use of UI elements. Always use components that are predictable and familiar.

EXAMPLE

The map in figure 7.4 shows the third-largest city measured by land area in the state of California. Without additional clues, a user would probably not be able to guess where or what this city is and be lost in the desert fog. The city is California City, incorporated in 1965 and master planned to rival other major cities in California. Located about 100 miles north of Los Angeles in the Mojave Desert, it has a current population of about 15,000.

To better communicate the content of the map, builders could add city labels and other key features such as highway shields. Another method to provide important context and identify the location faster is an **overview map**. Adding a **home button** or **location finder** can be helpful to reset the current map view.

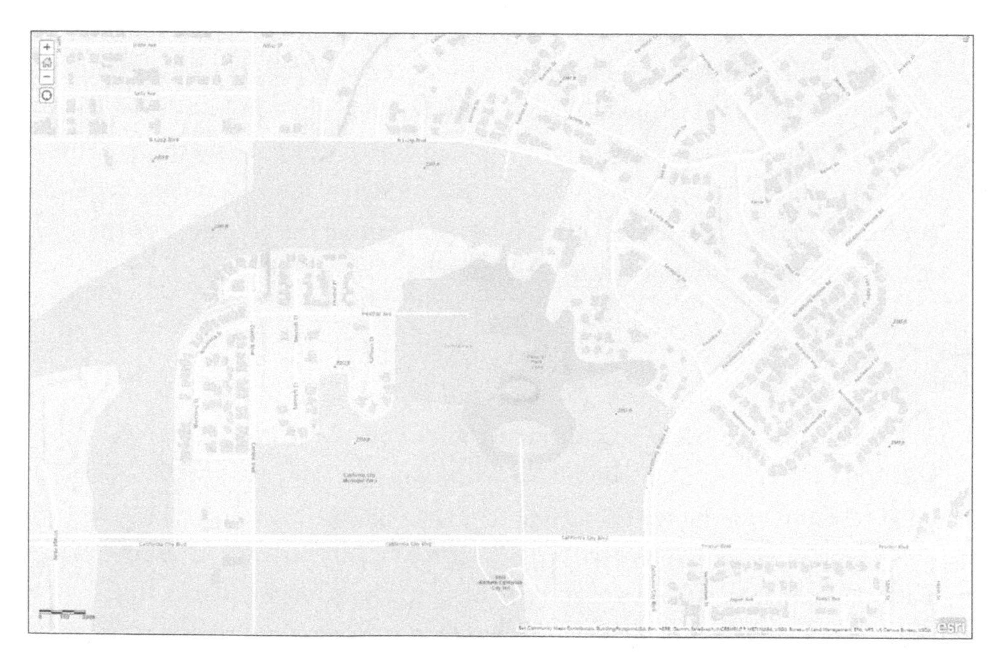

Figure 7.4. Without clues to orient them, users are lost in the desert fog.

Data–ink ratio

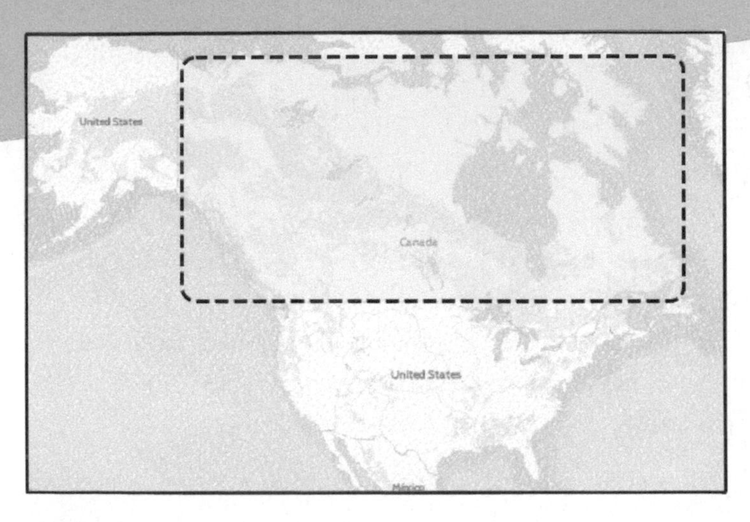

Figure 7.5. Data–ink ratio pattern.

WHAT

The **data–ink ratio** pattern (figure 7.5) relates to the amount of ink used to provide useful information. It compares the amount of data versus the amount of ink, or pixels, used in the display. A well-designed app should aim for a high ratio of relevant to irrelevant information, or a high ratio of data to ink. The goal is to increase the ink used to communicate signal and reduce content that contributes to noise. Following this rule will improve the efficiency of the map in communicating necessary information.

WHY

At a time when big data is omnipresent, it becomes more important to edit and curate the content and its visualization. The American statistician Edward Tufte uses the term *data–ink ratio* to argue against using excessive decoration in visual displays and calls superfluous information "chart chunk." Tufte states that an effective graph should aim to maximize the data–ink ratio, which represents the ratio of relevant to irrelevant information. For a well-designed map, this means that a large share of the pixels should present useful information.

WHEN

As a basic rule, apply the data–ink ratio to any map that is part of an app. Pay special attention to single-purpose apps (see chapter 6) that are designed for people with less GIS experience and little time and patience to work through their task. These apps require that the map and the UI components are built with minimum distraction and chart chunk. Data–ink ratio is also vital for maps on mobile devices that have limited space to paint needed information. Any noise on a small screen will severely decrease usability.

HOW

Always strive to show what is important while simultaneously removing what is not important. Start by asking, "What is the minimum set of visuals necessary to communicate the information understandably?" and continue by asking, "Which elements, such as layers, features, labels, or ornaments, can be removed without degrading the essence that needs to be communicated?"

Increase the amount of pertinent data points and decrease unnecessary content. The following techniques help increase the data–ink ratio:

- **Basemap layer:** Choose simple basemap layers such as light gray, dark gray, or human geography by default. Use **basemap toggle** to change the basemap layer to a visually more challenging option such as Imagery.
- **Operational layers:** Remove or hide layers that aren't necessary. Use **layer list** or **theme toggle** to make selected layers visible. Simplify features and apply scale dependency that hides layers when they are not needed at a certain scale.
- **Labels:** Be selective and show labels only when they help the reader. Apply scale dependency, and decrease label density. Prefer sans serif fonts over serif fonts.
- **Cartography:** Follow cartographic rules such as reduced color choices and strong contrast.

Another common problem related to the data–ink ratio is the display of non-continuous geographies. Noncontinuous geographies are areas that are spread apart—for instance, the US states of Alaska and Hawaii are geographically dispersed from the continental states. As shown in figure 7.5, showing all 50 states at the same time would also require displaying Canada, which uses most of the ink but isn't relevant in the context of the US-centric map. Use the following solutions to work around the problem of noncontinuous geographies:

- **Placemarks:** Use shortcuts to predefined places such as smaller territories to avoid showing widely dispersed data.
- **Inset maps:** Use multiple views to move dispersed geographies closer to each other.

EXAMPLE

A common problem in apps that show data for the US is that the states of Alaska and Hawaii, as well as US territories, are widely dispersed. Builders must make the difficult decision to default the map extent on the continental 48 states or show all. Showing all states and territories will decrease the data–ink ratio, whereas the former fails to be inclusive. The app in figure 7.6 uses inset maps to display Alaska, Hawaii, and the Virgin Islands close to the continental states. Each inset view makes intelligent use of local map projections to preserve area and shape. **Info pop-ups** are moved out of the small view window and attached to the main app. Because Alaska is a large state, it is important to let users navigate its view. On the contrary, there isn't much value in zooming or panning the views of Hawaii and the Virgin Islands, so their navigation is disabled. Inset maps are difficult to display on mobile devices, so the app removes them on small-screen resolutions.

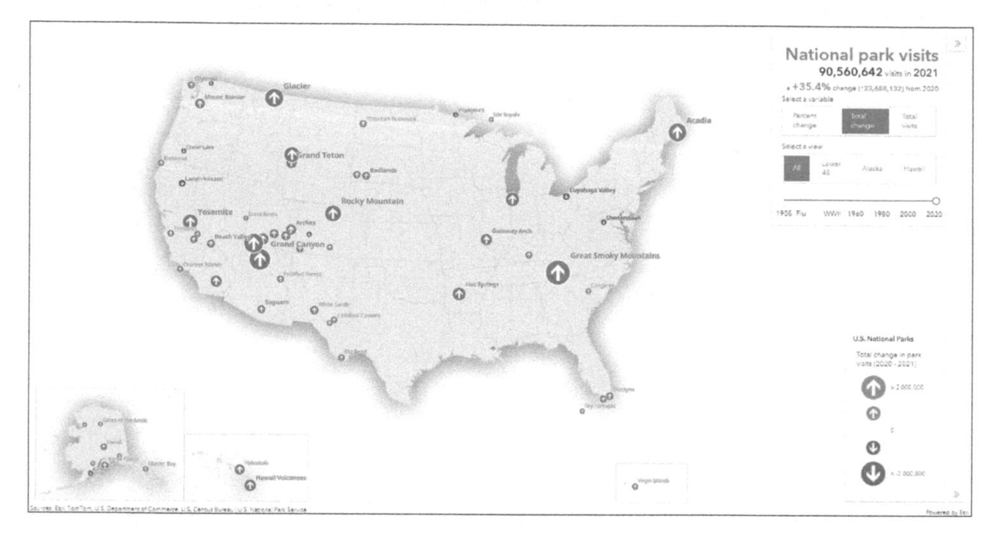

Figure 7.6. App uses inset maps to show change in visits to US national parks across the continental states, Alaska, Hawaii, and the Virgin Islands.

References

Chen, X. 2018. "Signal-to-Noise Ratio." September 9, 2018. www.nngroup.com/articles/signal-noise-ratio.

Ekenes, K. 2021. "Inset Maps for the Web." *ArcGIS Blog*, May 3, 2021. www.esri.com/arcgis-blog/products/js-api-arcgis/mapping/inset-maps-for-the-web.

Nielsen, J. 1994. "10 Usability Heuristics for User Interface Design." Updated Nov. 15, 2020. www.nngroup.com/articles/ten-usability-heuristics.

Tufte, E. 1983. *The Visual Display of Quantitative Information.* www.edwardtufte.com/tufte/books_vdqi.

About Esri Press

At Esri Press, our mission is to inform, inspire, and teach professionals, students, educators, and the public about GIS by developing print and digital publications. Our goal is to increase the adoption of ArcGIS and to support the vision and brand of Esri. We strive to be the leader in publishing great GIS books, and we are dedicated to improving the work and lives of our global community of users, authors, and colleagues.

Acquisitions

Stacy Krieg
Claudia Naber
Alycia Tornetta
Craig Carpenter
Jenefer Shute

Editorial

Carolyn Schatz
Mark Henry
David Oberman

Production

Monica McGregor
Victoria Roberts

Sales & Marketing

Eric Kettunen
Sasha Gallardo
Beth Bauler

Contributors

Christian Harder
Matt Artz
Keith Mann

Business

Catherine Ortiz
Jon Carter
Jason Childs

For information on Esri Press books and resources, visit our website at esri.com/en-us/esri-press.